Routledge Introductions to Development

Series Editors:
John Bale and David Drakakis-Smith

Latin America

Latin America represents one of the richest and most complex regions in the world in terms of its culture, history and politics. Some of Latin America's most enduring characteristics – language, religion, land ownership, industrial patterns and social inequality – can only be understood in the context of its colonial legacy. The countries of the region are currently suffering, both economically and socially, from the debt crisis. The question of their recovery is of international concern.

In this book, Alan Gilbert describes the development of Latin America from colonization in the sixteenth century to the present day. He interprets both its progress and its poverty as the outcome of external influences and internal dynamics. The book covers a wide range of issues – agricultural development, rural conditions, land tenure, reform and colonization, industrialization, the debt crisis, income distribution, employment and unemployment, migration, urban housing and services, class, race and gender, and the nature of Latin American politics – setting them firmly in their historical context.

The book will prove an invaluable introduction to Latin America for students of development studies, geography, politics, history, economics and sociology.

A volume in the **Routledge Introductions to Development** series edited by John Bale and David Drakakis-Smith.

In the same series

John Cole
*Development and Underdevelopment
A Profile of the Third World*
David Drakakis-Smith
The Third World City
Allan and Anne Findlay
Population and Development in the Third World
Avijit Gupta
Ecology and Development in the Third World
John Lea
Tourism and Development in the Third World
John Soussan
Primary Resources and Energy in the Third World
Chris Dixon
Rural Development in the Third World

Forthcoming

Rajesh Chandra
Industrialization and Development in the Third World
Graeme Hugo
Population Movements and the Third World
Janet Momsen
Women and Development in the Third World
Peter Rimmer
Transport Patterns in the Third World
Steve Williams
*Global Interdependence
Trade, Aid, and Technology Transfer*
Joe Doherty
The Socialist Third World
David Drakakis-Smith
Southeast Asia
Allan Findlay
The Arab World
Ronan Paddison
Retail Patterns in the Third World
Alan Jenkins and Terry Cannon
China
Tony Binns
Tropical Africa

Alan Gilbert

Latin America

London and New York

First published 1990
by Routledge
11 New Fetter Lane, London EC4P 4EE

Simultaneously published in the USA and Canada
by Routledge
a division of Routledge, Chapman and Hall, Inc.
29 West 35th Street, New York, NY 10001

© 1990 Alan Gilbert

Printed and bound in Great Britain by
Biddles Ltd, Guildford and King's Lynn

All rights reserved. No part of this book may be reprinted or
reproduced or utilized in any form or by any electronic,
mechanical, or other means, now known or hereafter invented,
including photocopying and recording, or in any information
storage or retrieval system, without permission in writing from
the publishers.

British Library Cataloguing in Publication Data

Gilbert, Alan, *1944 Oct. 1*
 Latin America. —
 (Routledge introductions to development)
 1. Latin America, history
 I. Title
 980
 ISBN 0-415-04199-6

Library of Congress Cataloging in Publication Data

Gilbert, Alan, 1944–
 Latin America/Alan Gilbert.
  p. cm. — (Routledge introductions to development)
 ISBN 0-415-04199-6
 1. Latin America — Economic conditions.
 2. Latin America — Social conditions.
 I. Title. II Series.
 HC123.G47 1990   89-48897
 330.98—dc20   CIP

Contents

List of tables	ix
List of figures	xi
List of plates	xiii

1 The physical and historical backcloth   1
  The colonial heritage   1
  Pre-conquest America   1
  Conquest and colonization   2
  The colonial legacy   3
  Independence   6
  Fuller integration into the world economy   8
  The physical backcloth   10
  Key ideas   12

2 Economic and social development during the twentieth century   13
  Economic and social change   13
  Case study A: The population explosion   16
  The export economy   18
  Import-substituting industrialization   20
  Export-oriented industrialization   23

The economic crisis of the 1980s 25
Case study B: The causes of recession 26
Key ideas 29

3 Rural development 30
Historical background 30
Land reform 32
Case study C: Land reform in Chile 34
Land colonization 35
Case study D: The Brazilian Amazon programme 36
The commercialization of agriculture 38
Key ideas 43

4 Migration and urban development 44
Urban growth 44
Migration 45
Natural increase versus migration in urban growth 49
The national settlement system 50
Policies to discourage metropolitan growth 54
Case study E: Migration to Bogotá 56
Case study F: The development of Brasília 57
Key ideas 59

5 Inside the city 60
The economy of the city 60
The formal sector 61
The informal sector 64
Living standards 65
Housing and services 65
Political activity and protest 70
Case study G: Earning a living in the informal sector 72
Case study H: Building a self-help home 74
Key ideas 75

6 Poverty, development and inequality 76
Social conditions 76
Race and ethnicity 78
Social class 79
Gender 81
The distribution of income 82

Case study I: The socialist model in Cuba	84
The development model	86
Case study J: The Brazilian economic miracle	88
The nature of the state	89
Key ideas	91

Further reading and review questions   **92**

Index   **95**

Tables

2.1	Development levels in selected Latin American countries, 1980s	15
2.2	Share of labour force in agriculture and manufacturing, 1950–80	15
A.1	Population of largest Latin American countries, 1930–85	16
A.2	Population growth rates by country, 1920–85	17
2.3	Major exports by country, 1985	19
2.4	Growth of industrial value added in major countries, 1950–85	22
2.5	Latin American export performance, 1965–85	25
2.6	External debt of selected Latin American countries, 1987	27
3.1	Major agrarian reforms in Latin America	32
3.2	Latin American agricultural production, 1960–86	41
4.1	Urban growth in Latin America, 1930–80	45
4.2	Native-born population as proportion of total population by age cohort, Caracas 1981	50
4.3	Urban primacy in selected Latin American countries	51
4.4	Population of Latin America's largest cities	53
5.1	Share of employment in different sectors, by city	61
5.2	Employment growth in manufacturing industry in selected countries	64
5.3	Housing conditions in selected cities	66

5.4 Relative growth of irregular settlement in selected cities 66
6.1 Social conditions by country, 1950–85 76
6.2 Open unemployment rates in selected cities 78
6.3 Secondary and university enrolment in selected countries, 1960–80 80
6.4 Female participation in the labour force by major world region, 1975 81
6.5 Inequality of income in Latin America 82
6.6 Real incomes by socio-economic group in Brazil, 1960–76 83
6.7 Distribution of income in Cuba, 1953–73 83
I.1 Economic growth rate in Cuba, 1962–88 84

Figures

1.1 Independence in Latin America 7
1.2 Physical relief and major rivers of Latin America 11
2.1 Latin America's economic growth during the twentieth century 14
2.2 Debt, inflation, interest rates and gross domestic product per capita, 1974–88 28
D.1 The Brazilian colonization of the Amazon and Centre-west 37
3.1 Cuba: changes in agrarian property 1959–61 39
4.1 Origin of migrants to Guadalajara, 1980 48
4.2 Major cities of Latin America, 1985 52
4.3 Location of Latin America's new cities 55
F.1 Map of Brasília – central area and satellite towns 57

Plates

1.1	Spanish colonial architecture – Mexico City	4
1.2	Portuguese colonial architecture – Ouro Prêto's main square	5
2.1	Early import substitution – the Volta Redonda steelworks in Brazil began operating in 1940	21
3.1	Commercial agriculture – tea pickers in Colombia	40
3.2	Deprivation in rural areas – a poor village in Caribbean Colombia	41
3.3	Market sellers in a small market town	42
4.1	Migrants tend to draw from the 15–40 year age group leaving the old behind in the rural areas	47
5.1	Luxury housing in Rio de Janeiro	62
5.2	Modern department store in Guadalajara	63
5.3	Employment at the rubbish dump in Mexico City	63
5.4	Self-help housing can improve through time – Bogotá	67
5.5	Flimsy housing – a land invasion on the fringes of Bogotá	67
5.6	Rental housing: a *vecindad* in a consolidated self-help settlement in Guadalajara	68
5.7	Unsuitable land is often used for housing construction – a house with a view in Tijuana	69
5.8	A water tanker filling the oil drums in a self-help settlement	70
G.1	The informal sector – bootblacks at work in central Guadalajara	72

6.1 Educational provision has improved, even in the rural areas 77
6.2 Modern values have diffused rapidly through Latin America, often mixing with more traditional beliefs: a roadside shrine 87

1
The physical and historical backcloth

The colonial heritage

Latin America cannot be understood without recognizing the importance of the legacy left by more than three centuries of Spanish and Portuguese rule. From the arrival of Christopher Columbus in 1492, the Spanish and Portuguese gradually incorporated large areas of the American continent into their overseas empires. Some of Latin America's most enduring characteristics, language, religious beliefs, patterns of land holding, export orientation and social inequality, were firmly established during that period. However much Latin America has changed as a result of roughly one and a half centuries of political independence, the colonial mark is indelibly printed on the area.

Pre-conquest America

There is much disagreement about how many aborigines lived in 'Latin' America when the Spanish and Portuguese arrived. One estimate is as low as 7.5 millions, another as high as 100 million. What is certain is that most of these indian peoples were concentrated in the mountain areas. The rest of the region was rather sparsely populated; in South America, most of the area east of the Andes contained only a sprinkling of hunters and gatherers. What is also clear is that relatively few of the indigenous peoples survived the onslaught of conquest and colonization. By 1650 war, disease and what was, effectively, slavery had cut the population to perhaps one-twentieth its pre-conquest size.

Among the peoples of pre-Columbian America were several who had established large empires. The Incas dominated much of the Andes, controlling most of Bolivia, northern Chile, Ecuador and Peru. Further north, the Aztecs dominated central Mexico, and the Mayas ruled much of south-east Mexico and Guatemala. These civilizations had developed highly organized societies and had conquered a series of neighbouring peoples. Each built impressive architectural monuments both to celebrate their own power and to placate the gods.

Conquest and colonization

The Spanish destroyed these civilizations. Their policy was to subjugate the indian peoples by superimposing their own religious and institutional structures on the societies that they found. Within a very brief period, a handful of *conquistadores* had overcome limited military opposition and destroyed the citadels of the existing civilizations. Tenochtitlán, the Aztec capital, was demolished and replaced by the Spanish viceregal capital Mexico City; the temples of the sun god were sacrificed to the cathedrals of catholicism. In Peru, the Inca capital Cuzco was effectively demolished, the remains being used as the foundations for new Spanish buildings.

The Spanish conquest had two main goals: to exploit the wealth of the conquered territories and to 'educate' the captured peoples in the sense of converting them into the catholic religion.

The first aim was easily accomplished. The indigenous civilizations already mined silver in considerable quantities, and current production was simply diverted to Spain. With the introduction of technological advances in processing the silver ores, the major mines at Potosí and Guanajuato became the jewels of the Spanish conquest. For more than three centuries, Spanish galleons would make the perilous journey across the Atlantic carrying silver to the mother country. Developing the region's other resources, particularly its agricultural potential, was more difficult. It involved the reorganization of indigenous production and the mobilization of the labour force. Agricultural change was gradually achieved by establishing the *encomienda* system. The *conquistadores* were given responsibility for, and power over, the local people. The indians retained the land but were required to work for the *encomendero*.

The latter was required to contribute one-fifth of the income to the Spanish Crown and to 'educate' the indigenous population. Beyond this, there were no controls over how he administered the *encomienda*.

The second aim, that of 'civilizing' the indigenous population, was achieved both through the *encomienda* system and through the establishment of churches. The Spanish built new towns and cities throughout the colony. In every town, a church occupied one side of the main square. Catholicism became well established among the minority of indigenous peoples who survived the dual thrusts of imported diseases and savage exploitation.

Portuguese rule was established with much less drama. There was no Portuguese equivalent to Hernán Cortés or Francisco Pizarro advancing on horseback to destroy much larger armies of hostile indians. Brazil is often said to have been more colonized than conquered, and the fighting which took place was mostly against the French or the Dutch, rather than with the indigenous peoples. Initially, the Portuguese traded with the indians, being mainly interested in the abundant red dyewood after which Brazil is named. Only gradually did they establish a colonial administration, mainly as a precaution against growing French influence. By 1549, however, Brazil was formally made part of the Portuguese Empire and a capital established in Bahia (today Salvador). A sparsely distributed Portuguese community produced cattle and sugar, and relied increasingly on imported slave labour to work the land.

From the very beginning, therefore, there were marked differences between Portuguese and Spanish America. However, there were also important variations emerging within the vast Spanish realm. Differences due to climate and relief, the presence or absence of precious metals, and variations in the numbers of indigenous peoples created contrasting economies and societies. The form of development in the silver-producing areas was very different to that in agricultural areas; agricultural organization in highland areas of dense indian population was very different to that in the sparsely populated plains of Argentina or the slave-based plantation systems established in the Caribbean. Over time, these differences were to become more marked.

The colonial legacy

While the variations between different parts of colonial America were very important, several critical legacies were imposed by both Spanish and Portuguese rule. These features still dominate many aspects of life in

4 Latin America

Plate 1.1 Spanish colonial architecture – Mexico City

the region and constitute the principal reason why we can sensibly refer to this area of the world as 'Latin' America.

Language and culture

Although many indigenous languages survive, most Latin Americans speak either Spanish or Portuguese. There are substantial numbers of Quechua, Aymara, Guaraní and Maya speaking peoples but these are still minorities compared to the vast majority who communicate in one of the two European languages. Similarly, Roman Catholicism is by far the most significant religion, even if indigenous beliefs and voodoo practices from Africa have been incorporated into religious practice in Brazil and parts of the Caribbean. In addition, many of the customs of Latin America are still clearly Spanish or Portuguese. Bullfighting remains popular in many places, bureaucratic and legal procedures resemble those of the Iberian peninsula, and many mannerisms and attitudes are a direct inheritance from the colonial past.

Population structure

The second legacy lies in the structure of the population. Latin America's

Plate 1.2 Portuguese colonial architecture – Ouro Prêto's main square

population is a mixture of three racial types: caucasian, indian and negroid. A great deal of racial mixing occurred during the colonial period, principally because relatively few Spanish or Portuguese women migrated to the new colonies. Today, the majority of the Mexican and the Andean population is *mestizo*, a cross between European and indian. In the Caribbean, and in much of Brazil, the dominant strand is *mulatto*, a cross between white and black. In a few places, too, an intermixture of black and indian is dominant. Spatial variations are very marked; most Argentines look very European, most Cubans negroid, most Guatemalans or Bolivians very indian. The racial structure is undeniably a major outcome of Spanish and Portuguese rule.

External orientation

Colonialism established important links between Latin America and Europe. By the late eighteenth century Latin America had developed what remains today its predominant trading pattern: Latin America exported minerals and agricultural products, and imported manufactures and services. In Brazil, the exports were predominantly cotton and sugar; in Spanish America mainly silver and gold. This dependence on 'primary' exports and manufactured imports is arguably a continuing cause of Latin America's 'underdevelopment' (see Chapter 2).

Social structure

While the period since independence has seen many changes in social structure, there are still clear signs of the colonial impact. Social class structure continues to be associated with skin colour (see Chapter 6). The ownership of rural land, while less significant today as a source of economic and political power, is still a vital ingredient in determining social class. Most Latin American countries remain profoundly unequal societies. Incomes are very unfairly distributed, with the poorest one-fifth of the population earning less than 3 per cent of household income in Brazil, Mexico, Peru and Venezuela (see Chapter 6).

Independence

Independence had been achieved in most of Latin America by 1830 (see Figure 1.1). Early revolts against Spanish rule in 1810 occurred in Mexico, Argentina, Chile and Venezuela, but were gradually suppressed. Nevertheless, the intellectual and emotional foundations of independence had been laid, and during the 1820s liberation movements broke out in Argentina, Venezuela and Mexico. By 1829 the efforts of San Martín, Bolívar, Iturbide and their allies had swept the Spanish out of most of the region. The Spanish retained control only in the Caribbean: in the Dominican Republic until 1844, and in Cuba until 1898.

The Portuguese presence continued longer partly because Napoleon's invasion of the Iberian peninsula in 1807 meant that the Portuguese Crown moved to Rio de Janeiro. Brazil continued as a monarchy until it was declared a republic in 1889. It is significant, however, that the whole of Latin America had achieved independence many years before most parts of Africa and Asia.

Initially, independence made little difference to the majority of the people. The independence movement represented more a political than a social revolution. Independence was organized and manipulated by the locally-born élites, not by the mass of the population. Dissatisfaction with the arbitrary rules laid down by Spain was the principal motivation behind the independence movement. Creole élites wanted more licence to develop export production and to trade directly with Britain.

Spanish America gradually broke up into a series of separate republics.

Historical and physical backcloth 7

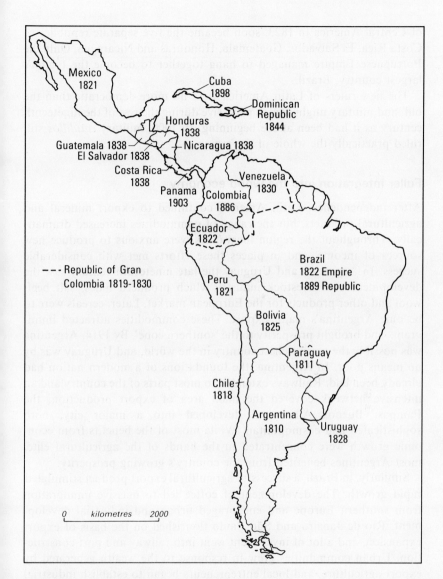

Figure 1.1 Independence in Latin America

The north-western republic of Gran Colombia had, by 1830, developed into three separate countries: Colombia, Ecuador, and Venezuela, and by 1903 a fourth, Panama. What had been declared as the United Provinces of Central America in 1823, soon became the five separate republics of Costa Rica, El Salvador, Guatemala, Honduras and Nicaragua. Only the Portuguese Empire managed to hang together to become the region's largest country, Brazil.

The new rulers of Latin America were no more democratic than the old, and military might was as pervasive during the rest of the nineteenth century as it had been at the beginning. In 1900, military *caudillos* still ruled practically the whole of the region.

Fuller integration into the world economy

After independence, Latin America continued to export mineral and agricultural products, but the range of commodities increased dramatically. Throughout the region local élites were anxious to produce new sources of income, and in places these efforts met with considerable success. In Argentina and Uruguay the late nineteenth century saw the development of a livestock industry which produced hides, salt beef, wool and other products for the European market. Later, cereals were to become Argentina's major export. These commodities attracted immigrants and brought prosperity to the 'southern cone'. By 1914, Argentina was possibly the tenth richest country in the world, and Uruguay was by no means poor. In Argentina, the foundations of a modern nation had already been laid. Railways extended to most parts of the country and an intensive network covered the main area of export production, the Pampas. Buenos Aires had developed into a major city, both sophisticated and cosmopolitan. While most of the benefits from economic growth were concentrated in the hands of the agricultural élite, most Argentines benefited from the country's growing prosperity.

Similarly, in Brazil, a successful agricultural export product stimulated rapid growth. The development of coffee led to massive immigration from southern Europe and encouraged urban and industrial development. Rio de Janeiro and São Paulo flourished on the basis of export expansion, and a lot of investment went into railway and port construction. Urban communities grew in response to the wealth generated by export agriculture, and local entrepreneurs began to establish industrial plants.

Elsewhere, however, strenuous efforts to develop exports were rewarded less generously. In Colombia, the first decades of independence produced a series of rather unsuccessful export 'booms' including gold, indigo, tobacco and cinchona bark (for quinine). In Venezuela, the situation was considerably worse, and Angostura bitters (which were drunk with gin) were for some time the major export! These differences in the ability to generate exports led to varying levels of economic development and very different kinds of society. Argentina was both rich and 'European', Bolivia remained poor and very indian. Argentina was industrializing and developing, and Buenos Aires became one of the world's largest and most affluent capitals. Bolivia continued to be a rural society; only a minority of workers were concerned with the production of tin for export. Neither agriculture nor tin produced much wealth for the majority; revenues from tin were monopolized by three companies, the 'tin barons', which dominated the Bolivian economy until the revolution of 1952.

The growing reliance on exports in most parts of Latin America meant an increase in what has often been called 'dependence' (see John Cole's book in this series). Many writers argue that this situation led to the 'developed' countries exploiting the 'less-developed'. That this has frequently occurred is undeniable. However, it is equally clear that some poor countries have become richer as a result of this process. Argentina, for example, was wholly dependent on markets in Europe and on capital from Britain, but managed to grow rapidly because of its dependence. Most Latin American countries are still dependent on developed countries today. In the nineteenth century attempts to raise the pace of economic growth invariably relied on increasing exports and thereby the level of dependence. What this produced within each Latin American economy, however, was highly variable. The production of bananas might result in a plantation economy dominated by overseas corporations; the export of copper was associated with highly sophisticated mining enterprises; coffee production sometimes encouraged the development of a relatively equitable system of small-scale agriculture; sugar production created a landed plantocracy which controlled the majority of the population.

Today, 'dependence' is no longer seen as an inevitable cause of poverty, it is viewed more as a condition which impedes progress and orientates development in ways which favour the interests of the developed countries. Poor Latin American countries have long been much more dependent on developed countries than the other way around. The problem is that the prosperity of the less-developed depends on the needs of the developed countries, on the latter's finance and technology. This situation has often led to the developed countries intervening in Latin America's internal affairs. Britain and, more recently, the United States

have actively interfered in Latin American politics. They have helped
carve out new countries (for example Panama and Uruguay), they have
displaced 'unfriendly' governments (in Chile and Guatemala), and they
have modified the development strategies of most Latin American
countries. But, even if the powerful nations of the world have clearly
influenced the nature of Latin American development, it would be
foolish to deny that some of the larger Latin American countries have
also had considerable autonomy from overseas control.

Most 'dependency' writers now place less emphasis on the exploitation
of Latin American countries by foreigners, and lay more stress on the
alliances that spring up between internal and external parties. It is these
alliances, say between a transnational banana producer and a military
dictator in Central America, or between the US State Department and
the president of a South American republic, which constitute the most
critical influence over the path of development. Total control over events
in, say, Paraguay is not in the hands of the United States alone. Rather,
such control rests within certain groups in the United States in alliance
with a minority of powerful interest groups in Paraguay. The form of
'dependence' is reflected in the nature of these class alliances; the
closeness of the ties by the nature and prosperity of export production
and by the degree of local autonomy. Dependence is a real problem but
its form is highly variable.

The physical backcloth

Throughout this summary of Latin America's past, the physical environment has emerged as a conditioning force in the region's human development. The distribution of the indigenous population was closely related to climate and physical geography, the growing of wheat in Argentina and bananas in Central America clearly linked to climate and soils. Obviously, Latin America cannot be understood without reference to its mountains and great rivers, its vast plains and forests. Clearly, the hugely varied physical geography of the region still influences the nature of life in each area. The developmental potential of windswept Patagonia is very different from either that of the Amazon jungle or that of the arid mountain plateaux of Bolivia. A flavour of this diversity is presented in Figure 1.2 which shows the major physical features of the region.

Something should also be said of the difficulties which natural disasters pose in Latin America. The western mountainous edge of the

Historical and physical backcloth 11

Figure 1.2 Physical relief and major rivers of Latin America

region, which includes all but six of the nineteen republics, suffers regularly from earthquakes and volcanic eruptions. The tragic events in Mexico City in 1985 merely repeated earlier tragedies in Managua in

1972, in northern Peru in 1941, 1962 and 1970, and in Chile in 1963. The weather also regularly endangers both human life and the economy. In and around the Caribbean, seasonal hurricanes destroy homes and crops. Drought affects considerable areas of the region, most frequently and seriously the north-east of Brazil. Frost attacks the coffee plantations of southern Brazil, and in many areas of the region heavy rains regularly cause floods and landslides in mountain and valley alike. Even the temporary shift of a sea current can devastate the local economy, for example the effects of *El Niño* on the fishing industry of Peru. While we will put little emphasis in this book on the physical environment, its effects on human development are always important and, sometimes, devastating.

Key ideas

1 Latin America was indelibly marked by Spanish and Portuguese conquest. Language, religion, social and racial structure, and patterns of trade were all transformed by three centuries of colonial rule.
2 Independence was achieved during the nineteenth century as a result of political, but not social, revolutions.
3 'Dependence' is an important factor in characterizing Latin American development, but its influence is highly variable and is not inevitably associated with poverty.
4 The physical background cannot be ignored. The effects on society of different kinds of climate and vegetation are of fundamental importance; the impact of natural disaster felt is all too frequently.

2
Economic and social development during the twentieth century

Economic and social change

Latin America's economy grew consistently from 1900 until 1980, except for a brief period during the 1930s (see Figure 2.1). The growth rate was quite slow until the 1950s, when the pace of change increased. Unfortunately, the debt crisis has meant that few Latin American economies have grown since 1980.

Severe though the consequences of the recent recession have been, they have not yet eliminated most of the benefits brought by almost eighty years of growth. Such a long period of economic growth brought genuine development to the region and raised living standards substantially above those typical in most parts of Africa and Asia. Of course, major differences are apparent within Latin America (see Table 2.1); Bolivia is rather poor but levels of development in Argentina, southern Brazil and Venezuela are not dissimilar to those of southern Europe.

The material gains from eighty years of economic development would undoubtedly have been greater had it not been for major improvements in life expectancy throughout the region. An increasing ability to control the most dangerous diseases produced dramatic results in Latin America; death rates began to fall during the 1930s and have continued to do so ever since (see Case study A). Throughout the post-war period these falling death rates, combined with more or less constant fertility levels,

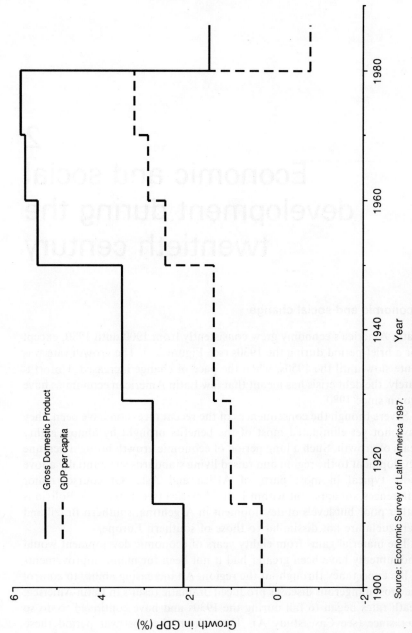

Figure 2.1 Latin America's economic growth during the twentieth century

Source: Economic Survey of Latin America 1987.

Table 2.1 Development levels in selected Latin American countries, 1980s

	Development indicator						
	1	2	3	4	5	6	7
Argentina	2,130	26	70	376	121	5	200
Bolivia	470	10	53	1,537	86	26	64
Brazil	1,640	26	65	683	107	22	127
Chile	1,430	27	70	1,231	109	6	115
Colombia	1,320	23	65	1,195	110	6	98
Cuba	n.a.	n.a.	77	486	124	n.a.	167
Dominican Rep.	790	15	64	1,763	103	23	90
Ecuador	1,160	20	66	1,043	90	18	64
Guatemala	1,250	9	60	2,184	100	45	27
Mexico	2,080	n.a.	67	1,035	127	10	111
Peru	1,010	18	59	1,111	92	15	49
Venezuela	3,080	36	70	816	108	13	125

Sources: ECLA (1987) *Anuario Estatístico para América Latina y el Caribe*, pp. 38, 42, 46, 50. World Bank (1987) *World Development Report*, pp. 202–3

1 Gross national product per capita 1985 in US dollars
2 Gross product relative to USA (USA=100), 1985
3 Life expectancy in years, 1985
4 Population per doctor, 1984
5 Calorie intake per person as percentage of minimum needs, 1981–83
6 Illiteracy rate among population aged 15 years and over, 1985
7 Television sets per thousand population, 1983

have stimulated rapid population growth. Between the 1950s and 1970s, for example, the populations of most Latin American countries were doubling every twenty-five years. Although rates of natural increase have now begun to fall, the average growth rate in many countries is still over 2 per cent per annum. There can be little doubt that the pace of

Table 2.2 Share of labour force in agriculture and manufacturing, 1950–80

	Agriculture		Manufacturing	
	1950	1980	1950	1980
Argentina	25.3	13.1	25.3	21.3
Brazil	59.7	30.4	12.9	17.9
Chile	36.2	16.3	19.4	16.7
Ecuador	64.4	34.8	10.1	12.5
Mexico	61.2	37.0	12.2	16.7
Peru	58.2	39.6	14.9	11.7
Venezuela	43.0	13.1	11.2	15.7

Source: J. Wilkie *et al.* (1988) *Statistical Abstract for Latin America* vol. 26

16 Latin America

Case study A

The population explosion

It is well known that the populations of most Latin American countries are growing very quickly. Since 1940, most countries have experienced periods when their populations were increasing by around 3 per cent annually. Improvements in health care and in general environmental conditions led to falls in the death rate. Since birth rates changed little, rates of natural increase rose, and most countries saw their populations increase rapidly. Table A.1 shows that the populations of Brazil, the Dominican Republic, Ecuador and Mexico increased by more than four times between 1930 and 1985, and that of Venezuela increased more than five times. Today, Brazil and Mexico are respectively the sixth and eleventh most populous countries in the world.

Only Argentina and Uruguay, and more recently Cuba, have had slower rates of population growth. The first two had experienced rapid growth during the early decades of the century and fertility rates had subsequently fallen. Cuba introduced a major programme of family planning during the 1970s which substantially reduced its growth rate.

Table A.1 Population of largest Latin American countries, 1930–1985 (millions)

	1930	1950	1970	1985
Latin America	102.0	158.5	272.9	389.1
Argentina	11.9	17.2	24.0	30.5
Bolivia	2.1	2.8	6.8	6.4
Brazil	33.6	53.4	95.8	135.6
Chile	4.4	6.1	9.5	12.1
Colombia	7.4	11.6	20.8	28.4
Cuba	3.8	5.9	8.6	10.1
Dominica Rep.	1.4	2.4	4.3	6.4
Ecuador	2.2	3.3	6.1	9.4
Guatemala	1.8	3.0	5.2	8.0
Mexico	16.6	27.4	51.2	78.8
Peru	5.7	7.6	13.2	18.6
Venezuela	3.0	5.0	10.6	17.3

Source: T. W. Merrick (1986) 'Population pressures in Latin America, *Population Bulletin*, vol. 41, p. 7, Population Reference Bureau. World Bank (1987) *World Development Report*, pp. 202–3. UNCHS (1987) *Global Report on Human Settlement 1986*, Table 1.
Note: Countries listed are those with more than 6 million inhabitants in 1985. The total for Latin America includes all twenty republics.

Case study A (continued)

However, the 1980s have seen a definite slowing in demographic growth throughout the region. Table A.2 shows that annual growth in the early 1980s was only 2.4 per cent compared to 2.9 per cent twenty years earlier. Only Bolivia and Venezuela failed to record a fall in the population growth rate.

Table A.2 Population growth rates by country, 1920-85

	Percentage annual growth			
	1920-25	1940-45	1960-65	1980-85
Total	1.9	2.2	2.9	2.4
Argentina	3.2	1.7	1.6	1.2
Bolivia	1.1	1.8	2.3	2.7
Brazil	2.1	2.3	2.9	2.3
Chile	1.5	1.5	2.5	1.7
Colombia	1.9	2.4	3.3	2.1
Cuba	2.7	1.6	2.1	0.6
Dominican Rep.	2.0	2.6	3.3	2.4
Educador	1.1	2.1	3.4	3.1
Guatemala	1.1	3.4	3.0	2.9
Mexico	1.0	2.9	3.5	2.9
Peru	1.5	1.8	3.1	2.8
Venezuela	1.9	2.8	3.3	3.3

Source: J. Wilkie et al. (1988) Statistical Abstract for Latin America, vol. 26, p. 109

The slowing growth rate has been due to families choosing to have fewer children. Birth rates in Colombia fell from 46 per 1,000 inhabitants in the early 1960s to 29 in the early 1980s. In Mexico, the fall was from 45 births per 1,000 to 30. The cheaper and wider availability of contraceptive devices has helped to bring about this dramatic decline. Despite opposition from the Catholic Church in some places, and pronatalist government policies in others (e.g. Argentina and Chile), family planning has been encouraged by private institutions in many countries. While few governments have actively encouraged family planning programmes, many have permitted large-scale private programmes. Surveys reveal that many women, especially in urban areas, are using some form of contraception.

Even with the rise in contraceptive use, however, the rate of national increase is unlikely to fall rapidly because of the age structure of the

Case study A (continued)

population. In most Latin American countries, more than half of the population is still under 15. As these children grow up they will have their own families, thereby sustaining the pace of demographic growth.

Rates of population expansion of over 3 per cent per annum have made the provision of adequate schools, jobs and services highly problematic. Both in rural and in urban areas, population pressure is growing. Increasingly, Latin American families recognize that a large number of children can be an economic burden.

Rapid population growth has hampered welfare improvement in Latin America. At the same time, it would be wrong to blame the persistence of poverty on the rate of population increase. After all, poverty among the poor has been a feature of Latin American society for centuries; it was a dominant characteristic even when population growth was very slow. In addition, there have been time when rapid population expansion has helped sustain economic growth, for example in Argentina, southern Brazil and Uruguay during the early years of the century. Population growth often makes the task of managing economic development more difficult. On the other hand, there are also many countries with slow population growth rates that have experienced little in way of economic development.

population growth has complicated the development process. It has increased rural population densities, it has led to rapid cityward migration, and it has worsened the already difficult task of providing services for the bulk of the population. In the process it has helped to transform a rather sparsely populated rural area into a predominantly urban region, turning mainly agricultural labour into a largely urban-based workforce (see Table 2.2).

The export economy

Throughout the twentieth century export production has been a key component in generating economic growth. Indeed, the development of each Latin American country is indelibly linked to the fortunes of one or two major export products. Brazil and Colombia are renowned for their coffee, Chile for its copper, Venezuela for its oil, Argentina and Uruguay for their meat and cereals. As is evident from this short list, Latin

Table 2.3 Major exports by country, 1985

Country	Product	% of total exports
Argentina	Wheat	13.5
Bolivia	Natural gas	59.8
Brazil*	Soya beans	9.9
Chile	Copper	46.1
Colombia	Coffee	50.2
Costa Rica	Coffee	32.2
Dominican Rep.	Sugar	25.9
Ecuador	Petroleum	62.8
El Salvador	Coffee	66.9
Guatemala	Coffee	42.5
Honduras	Bananas	31.1
Mexico	Petroleum	56.6
Nicaragua	Cotton	34.1
Panama	Bananas	23.3
Paraguay	Cotton	48.9
Peru	Copper	15.6
Uruguay	Wool	19.2
Venezuela	Petroleum	84.3

Source: Adapted from J. Wilkie *et al.* (1988) *Statistical Abstract for Latin America*, vol. 26, p. 432
*1985 was an exceptionally poor year for coffee production in Brazil, normally the country's major export.

America's exports have been dominated by agricultural or mineral products; until very recently Latin America exported very little in the way of manufactured exports (see Table 2.3).

In places, as we have seen, the successful development of export products brought a measure of prosperity as early as the nineteenth century. Coffee, for example, turned the south-east of Brazil into the main dynamo of the Brazilian economy. Spreading gradually westwards from Rio de Janeiro, the coffee economy brought prosperity in its wake. Brazilian entrepreneurs opened up large coffee plantations and imported workers from southern Europe to supplement local labour supplies. Money poured into the region to build railways and other forms of infrastructure. As the economy flourished local businessmen began to manufacture basic consumer goods. The southern Brazilian economy had begun a genuine process of development on the basis of export production.

Similarly, in Argentina and Uruguay a series of agricultural export products including mutton, wool, hides, wheat, alfalfa and beef transformed those nations into relatively wealthy countries.

In more recent times, the export that has produced the most dramatic change is oil. Venezuela, a backwater in the world economy until oil was discovered there in the early 1920s, was transformed by the new source of wealth. By 1928 Venezuela had become the world's second largest oil producer and its largest exporter. Within a few years, the Venezuelan economy had become very prosperous. In Mexico, large new discoveries in the 1970s stimulated a major boom in that country; lesser discoveries in Bolivia, Ecuador and Peru helped support otherwise-ailing economies.

Import-substituting industrialization

Until 1929, economic growth in most of Latin America was almost entirely linked to the fortunes of export production. Some industrialization had occurred in the more advanced countries to the south, but the basic model of development was similar to that which had been followed throughout the nineteenth century; economic growth depended on the generation of more agricultural and mineral exports.

This model came under direct threat in 1929, when the Great Crash signalled the beginning of a major world recession. Falling demand for their products in the industrialized countries led to a significant decline in Latin American export revenues. Since independence, the extreme reliance of Latin American countries on primary exports had always been underlined whenever there was a world recession; the recession of the 1930s was especially severe and marked a major change in the direction of Latin America's development.

In order to guard against too extreme a recession, certain governments introduced policies to maintain the level of internal demand. In Brazil, the government bought up most of the surplus coffee production, thereby sustaining the incomes of coffee producers and preventing the onset of an economic recession. In the absence of foreign exchange, however, it was difficult to import manufactured goods; local businessmen responded to the opportunity and began to produce substitutes.

Home-produced goods began to replace imported manufactures in the rest of Latin America, and by the late 1930s several governments were actively encouraging industrial development. During the Second World War, when demand for Latin America's agricultural and mineral products rose but imported manufactures were scarce, some of the unsatisfied demand was filled by home-made products. As a result, the share of industrial production in most Latin American countries rose: between 1929 and 1947, for example, industrial value added rose from 14

Economic and social development 21

Plate 2.1 Early import substitution – the Volta Redonda steelworks in Brazil began operating in 1940

per cent of Mexico's gross national product to 20 per cent, and in Argentina from 23 per cent to 31 per cent.

Following the end of the war, most Latin American governments formulated clear policies to foster 'import-substituting industrialization' (ISI). They improved the region's transport system and expanded the supply of electricity and water. They helped finance local industry and welcomed foreign corporations willing to establish factories. As a result, manufacturing production began to grow rapidly (see Table 2.4). Industrial employment rose dramatically and national self-sufficiency was achieved in most consumer goods. By 1960, all of the larger Latin American countries were producing their own textiles, processed foodstuffs, refrigerators and paint. Major steelworks had been established in Argentina, Brazil, Chile, Colombia, Mexico, Peru and Venezuela. By 1970, the larger countries were all assembling their own cars.

Unfortunately, in other respects the success of ISI was at best partial. First, ISI failed to resolve Latin America's tendency to import more than it could export. Indeed, ISI contributed to the problem because the new factories were dependent on foreign suppliers for machines, spares and intermediate products. Vehicle-assembly plants, established by North

Table 2.4 Growth of industrial value added in major countries, 1950–85 (per cent per annum)

	1950–60	1960–70	1970–75	1975–80
Argentina	4.1	5.6	3.4	–0.2
Bolivia	–0.4	7.9	6.8	4.7
Brazil	9.1	6.9	11.0	7.4
Colombia	6.5	6.0	7.8	3.4
Chile	4.7	5.3	–4.9	7.6
Ecuador	4.7	6.0	11.6	8.4
Mexico	6.2	9.1	7.1	7.2
Peru	8.0	5.8	5.7	1.1
Venezuela	10.0	6.7	5.2	5.1
Latin America	6.4	6.9	7.0	5.7

Source: Economic Commission for Latin America (UNECLA) (1986) *Anuario Estadístico de América Latina y el Caribe*

American and European companies, relied on supplies from Detroit, Paris and Hanover. And, since the prices of industrial components were usually higher than when they had been included in a finished, imported car, national import bills often rose. In addition, it was no longer easy to stem the flow of imports during a balance of payments crisis. If, for instance, the flow of vehicle parts was stopped, production in the assembly plants would be slowed and local jobs threatened. If economic growth was to be sustained at home, then foreign debts had to be incurred.

Second, levels of efficiency in the new industrial plants were sometimes rather low. At times this was a result of the limited size of the home market. Most production lines used technology designed to produce at much higher volumes than most Latin American economies could support. As a result, many plants were working at rather less than half capacity. In addition, many governments over-protected the new industries against overseas competition. High import tariffs on finished products, sometimes even quotas or total bans on imports, meant that the new factories were often given monopoly status. There was little incentive to improve management and labour practices and the prices of local manufactured goods rose relative to international prices.

Third, the rate of job creation was much lower than had been hoped for. In part, the problem was caused by the rapid growth of the labour force; too many new workers were demanding jobs as a result of high rates of population increase and rapid cityward migration. The problem was made worse, however, by the machinery used in the new industry;

rates of population increase and rapid cityward migration. The problem was made worse, however, by the machinery used in the new industry; capital-intensive technology was often used even when more labour-intensive methods were available. Managers sometimes preferred machines to men; although the former might occasionally break down, they were unlikely to go on strike. Legislation which was intended to improve labour conditions sometimes encouraged companies to hold back on recruitment. Whatever the precise causes, the numbers of jobs increased less rapidly than the numbers of workers available.

Finally, the process of ISI was bound to slow down eventually. In the smaller countries, it was only possible to establish plants producing a small range of products. It was inconceivable for Bolivia, Honduras, Nicaragua or Paraguay to contemplate building a car plant or steel works. Even in the larger countries, the process of ISI was destined, sooner or later, to run out of steam. Once the major consumer goods were being manufactured at home, other sources of expansion had to be sought. By the early 1960s, Argentina, Brazil and Mexico had achieved most feasible forms of ISI. A new approach was required.

Export-oriented industrialization

There were in fact two obvious, if difficult, alternatives. The first was to change the distribution of income within Latin America in order to modify the structure of demand. If more money was given to the poor they would consume more processed food, more textiles, and other home-made manufactures. This would stimulate local industry without attracting luxury imports. However, such a scheme would have been very unpopular among the better off and political realities meant that this approach was rarely tried. The second alternative was to export more in the way of manufactured goods. Of course, such an approach was anything but easy: since most of the technology and the inputs came from developed countries, it was difficult to see how the relatively inefficient manufacturing industries of Latin America would be able to compete. In any case, many of the companies were subsidiaries of transnational corporations; more exports from Latin America would mean more competition for the transnationals at home. Most 'experts' advised against this approach.

There was, however, a feasible variation on the second strategy. If it was too difficult exporting to the developed countries, why not export more to one another? If there was greater rationalization of production within Latin America, each company could produce for a larger market.

Brazil could produce one type of car, Argentina a second and Colombia a third; one kind of specialist steel could be made in one country, another kind elsewhere. Providing that each country was prepared to import the manufactures of its neighbours, every country would benefit. Free trade areas and common markets began to appear in the late 1950s. A multinational free-trade agreement was established between El Salvador, Guatemala, Honduras and Nicaragua in 1958 and two years later the group was joined by Costa Rica, establishing the Central American Common Market (CACM). The Latin American Free Trade Area (LAFTA) was set up in the same year, most of the larger countries soon becoming members. Unfortunately, neither group was very successful. Both faced the common difficulty that since the largest countries contained the most efficient industries, they reaped most of the benefits. While some compensation was offered to the poorer countries, the benefits were too concentrated. In addition, political differences between the countries sometimes led to major disagreements, the most spectacular conflict being the war that broke out after an international football match between El Salvador and Honduras.

In the light of this experience, the export of manufactures did not seem to be a serious option. However, during the 1960s there was an intellectual return to free trade thinking and attempts were made to encourage Latin American countries to export more to the developed countries. The stimulus here was undoubtedly the example set by the 'Gang of Four' (Hong Kong, Korea, Singapore and Taiwan), countries which were successfully penetrating the markets of the developed countries. Encouraged by advice from the World Bank, several governments including those of Brazil and Colombia began to reduce levels of domestic protection and to give incentives to export producers. By the 1970s this had become the accepted way of sustaining industrial expansion. And, as Table 2.5 shows, the policy appears to have had some effect. In Brazil, Colombia, and the Dominican Republic, the share of manufactured exports has risen impressively, and along the USA–Mexican border a series of *maquiladoras* (assembly plants) has been established in which cheap Mexican labour puts together imported materials for export to the United States.

Few believe, however, that all Latin American countries will be as successful as the new exporters of the Far East and southern Europe. For a start, most successful industrial exporters are located strategically close to developed countries: South Korea is close to Japan, and Spain to western Europe. Whilst Mexico and the Dominican Republic are well

Table 2.5 Latin American export performance, 1965–85

	Exports as percentage of gross domestic product		Manufactures as percentage of total merchandise exports	
	1961	1985	1965	1985
Argentina	8	15	6	16
Bolivia	21	18	4	7
Brazil	8	14	8	41
Chile	14	29	4	7
Colombia	11	15	7	18
Dominican Rep.	15	28	2	24
Ecuador	16	27	2	1
Guatemala	17	19	14	25
Mexico	9	16	16	27
Peru	16	22	1	12
Venezuela	31	27	2	5
Low-income countries	7	10	n.a.	44
Middle-income countries	17	26	20	41
Developed countries	12	18	70	76

Source: World Bank (1987) *World Development Report*, pp. 210–11, 222–23

located in this respect, most of the other countries of South America are distant from the United States, Europe or Japan. In addition, the existence of the 'newly industrialized countries' of Asia and southern Europe makes Latin America's task much harder. Of course, there are major opportunities for the export of more manufactures, but few expect this strategy to provide more than some of the answers to Latin America's development problems. After all, most Latin American countries are still reliant on primary exports (see Table 2.3).

The economic crisis of the 1980s

No sooner had export-oriented industrialization become a solid plank in Latin America's development strategy, than the debt crisis struck. Several Latin American countries had borrowed heavily in order to increase their export capacity and, just as the new investments were coming on stream, the world economy entered a recession.

The origins of the debt crisis are explained in Case study B; the problem lay now in how to repay the debt. In general, Latin American countries have endeavoured to repay as much as they can. Unfortunately, as Table 2.6 shows, interest repayments constitute on average nearly one-third of their export earnings.

26   Latin America

Case study B

The causes of recession

When the Organization of Petrol Exporting Countries (OPEC) was established in 1973, oil prices rose roughly four times. This immediately benefited Venezuela and, later, those countries such as Bolivia, Ecuador, Mexico, and Peru, which had found oil in substantial quantities after the price had risen. If higher prices helped the oil producers, however, they also brought major balance of payment problems for large oil importers such as Brazil and Argentina.

The rise in oil prices led to a world recession. The recession reduced the demand for products from less developed countries, a phenomenon which became more marked once the war in Vietnam ended. The worst effects of the recession were avoided, however, by the fact that the oil states of the Middle East began to spend heavily on imports and to invest their surplus revenues in Western banks. The rise of the so-called petrodollar money market put huge sums at the disposal of the Western banks.

Having accumulated so much money, however, it was incumbent upon these banks to lend it. As such, they were anxious to satisfy requests from Latin American countries for loans to overcome temporary balance of payments problems. Since interest rates were lower than inflation rates in the developed countries at this time, borrowing was effectively free. As a result, Latin American countries were happy to borrow large sums of money. Even petrol-rich countries such as Mexico and Venezuela borrowed extensively so that they could step up the pace of their development.

The borrowing strategy might have worked but for two developments. First, another rise in oil prices in 1979 led to a deeper recession in the world economy, which meant that Latin America's expanded manufacturing and mineral base was faced with a declining international market. Second, the introduction of monetarist financial policies in Britain, Japan and the United States pushed up interest rates. The cost of borrowing shot up suddenly and Latin American countries were faced by a major increase in their debt repayments.

The impact on Latin America was severe. For the region as a whole the ratio of interest payments to exports of goods and services rose from 9 per cent in 1974 to 41 per cent eight years later. In 1982, Argentina and

Economic and social development 27

Case study B (*continued*)

Brazil faced interest/export ratios of well over 50 per cent, way beyond their ability to pay.

A major financial crisis threatened, and several banks in the developed countries faced bankruptcy if their major Latin American debtors defaulted. As such, there was considerable diplomatic pressure on Latin America to pay at least the interest on the loans. While there were suggestions, particularly from Fidel Castro, that Latin Americans should refuse to repay, no debtors' club was ever formed. Apart from Peru, which refused to commit more than ten per cent of its export revenues to interest payments, most Latin American countries stepped up their debt repayments.

The economic and social consequences of these attempts to repay the debt have been profound. Most Latin American countries have had to deflate their economies and cut imports. Government expenditure has been severely pruned and subsidies to the poor reduced. Local currencies have been devalued, stimulating inflation at home. Rates of inflation have been far higher than wage increases; in most countries wage earners have suffered very badly.

Table 2.6 External debt of selected Latin American countries, 1987

	Total debt (US $ billions)	Debt per capita (US $ billions)	Interest due as percentage of export earnings 1987
Latin America	410.5	1006	29.7
Argentina	56.8	1818	51.0
Bolivia	3.9	578	43.9
Brazil	114.6	808	33.1
Chile	19.1	1526	26.4
Colombia	15.9	537	20.7
Dominican Rep.	3.8	566	14.7
Ecuador	10.5	1051	32.8
Guatemala	2.8	331	13.6
Mexico	96.7	1159	29.8
Peru	16.2	824	21.9
Venezuela	31.9	1728	23.7

Source: UNECLA (1988) *Anuario Estadístico para América Latina y el Caribe*, pp. 24–25

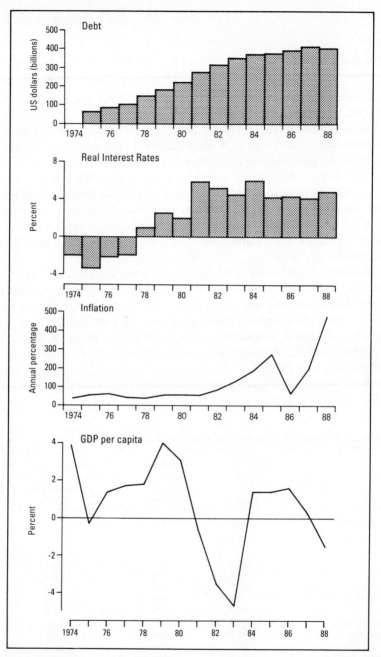

Figure 2.2. Debt, inflation, interest rates and gross domestic product per capita, 1974–88

Economic and social development   29

In attempting to repay the debt or to maintain interest payments, countries have squeezed demand at home. The pace of economic growth has plummeted. Between 1981 and 1989, per capita gross domestic product in the region fell by 8 per cent, it increased in only four out of twenty countries. Living standards have fallen in most countries. In Argentina, per capita gross domestic product in 1985 was the equivalent of what it had been in 1968, in Bolivia it had fallen to the level of 1966, in Peru to that of 1962 and in Venezuela to the level of 1963. No wonder poverty levels have increased.

Key ideas

1 With the exception of a brief interlude during the 1930s, most Latin American economies grew from 1900 until 1980.
2 Exports led economic growth until the 1930s when a phase of import-substituting industrialization began.
3 Import substitution generated industrial growth but not without a series of problems linked to inefficiency and low rates of labour absorption.
4 Since the 1960s, export orientation has again become a principal source of economic development.
5 Since 1980, the economic situation has been badly affected by the debt crisis.

3
Rural development

Historical background

The Spanish Crown rewarded the *conquistadores* in one of two ways: through placing them in charge of an *encomienda* or by granting them ownership of a large tract of land. Under the first system, the *encomienda* was given control over a group of indians from whom he could extract work or tribute. In exchange for this right, he was expected to give approximately one-fifth of his income to the Spanish Crown. In addition, he was required to maintain law and and order and to convert the local population could work the land and was prepared to pay tribute to the new masters. It worked less well where the indians were less docile, for example in Agentina and in the south of Chile, and in sparsely populated areas. Under such conditions, and throughout most of Portuguese America, land grants were the commonest method of reward, a single individual sometimes controlling several thousand hectares.

Both systems led to a highly concentrated land-tenure pattern. The difficulty was how best to extract work from the reluctant indian labour force. Over the years, a whole range of different methods evolved. In places, landlords encouraged the indians to get into debt and then forced them to pay off their obligations through work. Elsewhere, a form of forced labour was used, the *repartimiento*. A further variation was share cropping, a system where land was given to the indians in return for a share of their harvest.

Such methods were supplemented in the tropical areas by slavery. The limited populations of Portuguese America and most Caribbean islands, together with the reluctance of many indians to work on the plantations, encouraged the imports of slaves from America. From the middle of the sixteenth century, the sugar and cotton plantations were almost entirely dependent on slave labour. Independence from Spain and Portugal brought little change. A few slaves were freed as a reward for their help in the independence movement but, where slavery was the major source of labour, the practice persisted longer; in Puerto Rico until 1878, in Cuba until 1886 and in Brazil until 1888.

Even when slavery was declared illegal, it continued informally; labour systems based on debt-peonage and share cropping became more common. In places, independence made the situation worse. In Mexico, for example, the abolition of corporate land holding in 1856 led to church and community lands being converted into freehold property. Lacking the means to buy the land, the villagers saw their communal property sold and concentrated into a series of large *haciendas*. Thenceforth, they were forced to work for the landowner.

In certain places, agricultural colonization sometimes allowed smallholders to obtain land. In Colombia, many small coffee farms were established by settlers in the west of the country. Even here, however, large landlords claimed most of the better land, a process that was to become still more apparent in the massive land colonization schemes of the twentieth century in other parts of South America.

Limited access to agricultural land gave rise to a series of practices which reduced the productivity, and hence the incomes, of the majority of rural inhabitants. In places, it led to the landless labour force being forced to work for low wages. Elsewhere, it meant that smallholders were forced to eke out a living on tiny plots, while the estate occupied most of the more fertile, flat land. The peasantry were sometimes given enough land to produce the food to keep themselves alive, but insufficient to deter them from working for wages. While different systems developed in different parts of Latin America, the common element stands out. By controlling land, the politically powerful were able to obtain a cheap labour force. Deprived of sufficient land themselves, the majority were forced to work others' land for low wages. All too often, it also meant that the local élite literally controlled the lives of the peasantry. When it came to elections, the peasantry voted for whoever the landowner nominated.

Table 3.1 Major agrarian reforms in Latin America

Country	Period of major activity	Nature of reform
Bolivia	1953–70	Large estates divided among sharecroppers, indigenous communities ties established. Worst forms of labour exploitation made illegal. Commercial farms protected.
Chile	1967–73	Expropriation of largest farms, particularly after 1970 under Allende's socialist government. Many land invasions of large estates 1970–73. Most of the reforms reversed after 1973 when military government takes power.
Cuba	1959–63	Major reform taking over the lands of foreign sugar estates and cattle farm companies to form state enterprises. Tenant farmers received some land.
El Salvador	1980	Division of some very large estates with some land going to former tenants.
Guatemala	1952–54	Major redistribution of land which was reversed when the Arbenz government was removed with the help of the United States.
Mexico	1917–89	The pioneer reform breaking up the largest *haciendas* and redistributing the land either as smallholdings or as *ejidos* (community-owned land). Most of the latter established during the administration of Lazaro Cardenas (1934–40). The reform has continued and by 1989 some 77 million hectares had been redistributed.
Nicaragua	1979	Expropriation of the lands of former dicator Somoza to form state farms. Some idle land transferred to tenants. Encouragement given to co-operatives.
Peru	1969–75	Expropriation of major sugar and cotton estates and establishment of worker co-operatives. *Haciendas* in the Andes turned into peasant communities and incipient co-operatives. In total, 35% of agricultural land expropriated. Reforms partially reversed after 1975.

Source: Adapted from Preston (ed.) (1987) *Latin American development*, Table 8.4

Land reform

Revolution and protest have occasionally brought changes to the pattern of landholding, and during the twentieth century several countries have experienced major agrarian reforms. Elsewhere, governments have approved legislation of varying effectiveness. Table 3.1 summarizes the most significant attempts at land reform in the region.

The earliest major reform was a direct outcome of the Mexican Revolution (1910–17). The annexation of communal lands during the nineteenth century, together with the attraction of foreign investment and the commercialization of farming under Porfirio Diaz (1876-1910), had led to an acute level of land concentration. It is commonly said that 1 per cent of the population owned 97 per cent of the land, while a further 96 per cent of the population controlled only 1 per cent. Whatever the accuracy of those figures, the vast majority of peasant families had no land.

The land reform of 1917 broke up the largest *haciendas* and distributed some of the land to the poor. Many families were given their own smallholdings. Unfortunately, there was insufficient land to give plots to every family and the government decided to resolve the problem by establishing areas of community land. These *ejidos* gave inalienable control to the indian communities, members of the community having the right to farm part of the land during their lifetime. The reform was neither total nor immediate; most *ejidos* were not established until the 1930s and the reform process has ebbed and flowed ever since. Despite the reforms, rural Mexico remains a very unequal place.

Major land redistributions have occurred only in a handful of other Latin American countries: notably in Bolivia, Chile, Cuba, Guatemala and Nicaragua. All of these reforms have been associated either with a revolution or with the election of a left-wing government, the latter event usually proving a disappointment because the military has quickly intervened to remove what they saw as the 'danger' of reform. In Guatemala in 1954, and in Chile in 1973, democratically elected governments were removed by the military, the latter receiving considerable assistance from the United States. In both countries, most of the redistributed land was given back to the original owners.

In Bolivia, Cuba and Nicaragua, revolutions led to the expropriation of the largest estates but in each case the form of redistribution was very different. In Cuba, the estates formerly controlled by foreign sugar and cattle corporations were transferred to the state and co-operative sector (see Figure 3.1). In Nicaragua, by contrast, many of the lands given up by the deposed dictator and his family were distributed to individual families.

Elsewhere, the effects of agrarian reform have been limited. Admittedly, the 1960s saw the United States government pressing most Latin American countries to introduce reformist legislation. The Alliance for Progress offered foreign aid in return for reform. It resulted in new laws being introduced in Venezuela in 1960, in Colombia in 1961, in Chile in 1962, in Brazil in 1963, and in Peru in 1964. While some land was redistributed, particularly in Chile and Venezuela, there was little real

Case study C

Land reform in Chile

The first land reform in Chile was approved in 1962 and modified in 1967. At the time it was seen as the pioneer of Alliance for Progress reformism in Latin America. But while some land was redistributed, progress was slow; between 1964 and 1970, the government of Eduardo Frei gave land to 30–37,000 families, many thousands less than the target of 100,000.

Salvador Allende took office in 1970, leading the first democratically elected Marxist government in Latin America. Rural expectations were naturally high and the government responded by accelerating the reform programme. The pace of expropriation was stepped up. The expropriated land was sometimes organized into co-operatives and sometimes into state farms; by the end of 1972, more than one-third of the country's agricultural land was in the reformed sector. In addition to the official programme, there was also a rash of land invasions, a process encouraged by the more radical parties in the socialist government coalition.

By 1973 it was claimed that private properties of over 80 hectares no longer existed in Chile, a marked contrast to the situation in 1965 when 55 per cent of the land was held in estates of that size. The reform had clearly been successful in dismantling the large estates.

Unfortunately, the general economic problems of Chile at the time, together with the threats being made by some of the more radical elements in the government, led to a dramatic fall in marketed agricultural production. Farmers fearing expropriation sold off their livestock and failed to plant new crops. Food shortages began to appear in the cities and the cost of food imports rose to one-third of the value of total exports in 1972. Chileans are still divided between those who regard the agrarian reform as a major success and those who believe the decline in agricultural production was an avoidable disaster.

It is impossible to say what might have happened in the Chilean countryside if socialist rule had continued, for the military intervened in September 1973. President Allende was killed and many of his supporters were murdered, jailed or exiled. General Pinochet soon reversed the land redistribution process. New expropriations ceased and, where land was deemed to have been expropriated illegally, it was returned to the

Case study C (continued)

original owners. Almost three million hectares, out of a total of almost ten million hectares, were given back. By 1977, the large estate had returned and there were now more than 1,500 estates over 80 hectares. Between 1973 and 1978, some 35,000 private owners received land titles, and the development of co-operatives was discouraged. Some of the remaining expropriated land was assigned to smallholders but much was retained by the reform agency because it was unsuitable for agricultural use.

change in the land-tenure pattern. Indeed, the Alliance for Progress reforms are widely regarded as a cosmetic exercise, and have even been called a 'counterreform' because they blocked efforts at real change.

Land colonization

In the absence of significant amounts of land redistribution, figures on agrarian reform were sometimes conflated with data on the opening up of new land in colonization areas. Either through the extension of irrigation systems or the cutting down of forest, large areas of land have been brought into production during the last fifty years.

In Mexico, the area of irrigated land has increased enormously during recent years. The building of a series of large dams has transformed farming in the major river basins of the country. In the 1950s there were approximately one million irrigated hectares; by the early 1980s, the Mexican government claimed that 3.2 million hectares of irrigated land had been harvested.

Further south, irrigation has been less important than deforestation in extending the cultivated area. The most spectacular effort, of course, has been in Brazil where the opening up of Amazonia and the mid-west has been a very ambitious, if highly flawed, enterprise (see Case study D). Substantial efforts have also been occurring on the northern and western edges of the Amazon as the Andean republics have sought to counter Brazilian colonization efforts close to their borders.

Case study D

The Brazilian Amazon programme

The Amazon region contains approximately one-twentieth of the world's land surface. Until recently, much of this area was simply a sparsely-populated tropical forest. Since the mid-1950s, however, successive Brazilian governments have made determined efforts to develop the region.

The first steps towards opening Brazil's interior were the building of Brasília and the construction of a road linking the new capital to Belém in the north. The latter was finished in 1964, although paving along the road's whole 2,276 kilometres was not completed until 1973. Construction of the road led to large numbers of farmers moving into the region.

After the military coup of 1964, the government gave high priority to the Amazon programme. A new development agency was established for the area and generous tax incentives were offered to companies willing to invest in the region. Major new roads were constructed, notably the Transamazônica which linked the north-western cities with the Peruvian border – a distance of some 3,500 kilometres. Between 1970 and 1974, the pace of construction was frenetic, the road network growing from 1,500 kilometres to 8,600 kilometres in just five years. The new roads were intended to attract settlers to the Amazon from deprived agricultural regions. They were also intended to tempt major companies into developing the area's mineral, timber and cattle-ranching potential.

The opening of the Amazon region has clearly been a major achievement. In 1980 some 13.3 million people lived in the states of north and central-west Brazil, compared with only 3.6 million in 1950. Mineral, beef and timber production have increased dramatically, providing a major contribution to export earnings. From a social and environmental perspective, however, the programme has been much less of a success. The schemes to attract poor agriculturalists from the drought-ridden north-east were plagued by a lack of both careful planning and resources. Those smallholders who migrated were faced by difficult physical conditions and rarely received title deeds to the lands they cleared. Many cut down the forest and planted crops only to be displaced by large corporations holding land title to vast swathes of territory. The settlers have been forced to move on and have come into increasingly brutal conflict with the indigenous peoples. The indian populations have

Case study D (*continued*)

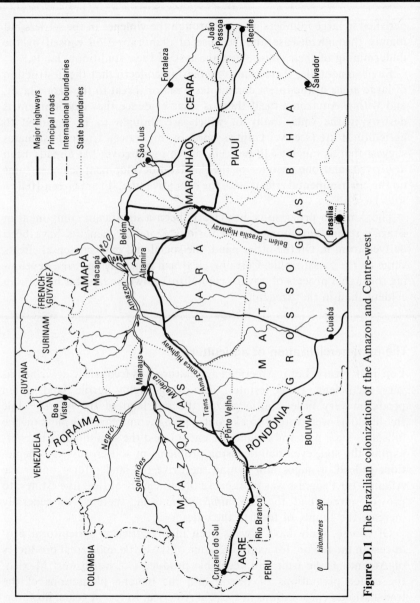

Figure D.1 The Brazilian colonization of the Amazon and Centre-west

Case study D (continued)

perished in large numbers, in part through the violence of the settlers but mainly through disease and because of demoralization caused by the undermining of their culture and the loss of their traditional lands.

Environmentally, there is also increasing concern that the destruction of large areas of the forest constitutes a major threat to the ozone layer, and within Amazonia itself there is clear evidence that development is destroying the soils, many of which are thought to be unsuited to agricultural use (see Avi Gupta's book in this series). Current estimates suggest that as much as 10 per cent of the forest cover has already been destroyed, and one account of the Amazon development schemes sums up the environmental effects with the evocative title: 'From Green Hell to Red Desert'!

The Amazon programme has clearly been a significant component in Brazil's recent development and various economic benefits have been derived from it. On the other hand, the social and environmental effects verge on the catastrophic. Certainly, if Brazil generally has been developing through a process of 'savage capitalism', nowhere has this been more evident than in the Amazon.

The commercialization of agriculture

For many years, Latin American agriculture was frequently described as very inefficient. Large estates, occupying highly fertile land, often produced very little. Cattle grazed the flat land of the estates while smallholders were forced to eke out a living from marginal plots on the hillsides. Such patterns of land use encouraged the smallholders to over-exploit the soil, eventually causing problems of soil erosion and depletion. Indeed, a major aim behind the drive for land reform under the Alliance for Progress was to improve the efficiency of the large estates, to provide more land for the *minifundista* and, thereby, to generally increase the supply of food to the cities.

Of course, there have always been highly efficient agricultural producers in the region: for example, some of the cattle and cereal producers of Argentina, the fruit and vegetable producers of north-west Mexico, the coffee plantations of Brazil, and the banana plantations of the transnational corporations in Central America. In recent years, however, the agricultural sector has been transformed in certain parts of the region

Rural development 39

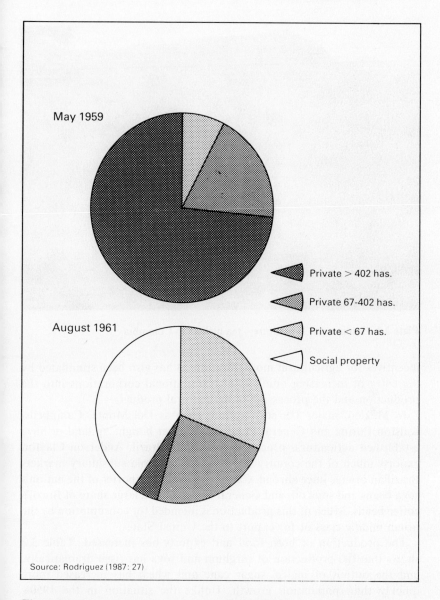

Figure 3.1 Cuba: changes in agrarian property 1959–61

by the development of commercial agriculture and agribusiness. This change has been encouraged by Latin American governments offering

40 Latin America

Plate 3.1 Commercial agriculture – tea pickers in Colombia

incentives for agricultural modernization. It has also been stimulated by the entry of increasing numbers of transnational corporations into the production and the processing of agricultural products.

In Mexico, major US corporations such as Del Monte, Campbells, Ralston Purina and General Foods have either bought up land or have established agricultural processing plants. In Brazil, Anderson Clayton exports much of the country's cotton, a Coca-Cola subsidiary markets Brazilian orange juice abroad, Cargill is a leading exporter of the nation's soya beans and soya oil, and General Foods buys a large share of Brazil's coffee beans. Much of this production is intended for consumption by the urban middle class or for export to the United States.

The production of both food and exports has increased. Table 3.2 shows that the production of sorghum and soya has risen dramatically, and the output of maize, sugar cane and wheat has increased more quickly than population growth. Unlike the situation in the 1950s, therefore, it would certainly be misleading to claim that all, or even most, of Latin American farmers are backward and unproductive. By the end of the 1980s relatively little subsistence production; most farmers are producing for the market and even for export.

Rural development   41

Plate 3.2 Deprivation in rural areas – a poor village in Caribbean Colombia

Whether this transformation has brought benefits to the majority of Latin Americans, however, is a moot point. First, agribusiness developments have changed food tastes in the region, particularly in the urban areas. There has been a shift away from maize-based staples such as *tortillas* and *arepas* toward bread. This has both increased the cost of a

Table 3.2 Latin American agricultural production, 1960–86 (thousand tons)

	1960	1986	Annual increase
Maize	22,422	52,819	3.4
Potatoes	8,506*	11,498	1.4
Rice	10,896	17,476	2.3
Seed cotton	3,486	4,483	1.0
Sorghum	2,369*	12,685	8.3
Soya	233	22,080	46.9
Sugar cane	235,789*	456,798	3.2
Wheat	8,016	21,581	3.9

Source: Economic Commission for Latin America *Anuario Estadístico de América Latina y el Caribe*, various years

*Figures for 1965

42 Latin America

Plate 3.3 Market sellers in a small market town

basic ingredient in the household budget and led to increasing imports of wheat. Mass advertising has encouraged the purchase of products with dubious nutritional value: cornflakes, ketchup and hamburgers. More expensive, less nutritious food is too often replacing traditional staples in the national diet. Second, many farmers no longer produce for the home market, having moved into the production of exports. As Burbach and Flynn (1980: 105) have commented: 'Land that once produced blackbeans in Brazil is now being used to grow the soybeans that are turned into the animal feed to fatten cattle for the Japanese market.' A basic item in the Brazilian diet is being displaced by indirect production for the export market. The situation became so serious in the early 1980s that there were food riots in São Paulo and attacks on supermarkets. The commercialization of agriculture has also accentuated the inequality of land holdings. In Brazil, large areas of the Amazon have been taken over by foreign companies producing for export; in Mexico large farmers are sometimes renting *ejido* land because the communities cannot afford to invest in essential fertilizers, tractors or irrigation systems.

One result of all these changes is that the Latin American countryside remains a deprived area. Living standards are generally lower than in the cities; most people are poor in the countryside. The average country dweller earns a low income, has little access to electricity or piped drinking water, and benefits little from education, health, and social security systems. This disparity between urban and rural living conditions continues to encourage many people to leave the countryside and migrate to the cities.

Key ideas

1 Spanish and Portuguese rule established a system of land tenure that was highly unequal.
2 During the twentieth century, several serious attempts at land reform have been made in number of Latin American countries. Despite these changes, the distribution of land continues to be very unequal in most countries.
3 Large areas of agricultural land have been opened up both through irrigation and through deforestation. Unfortunately, colonization has too often led to powerful farmers gaining control over vast tracts of land.
4 Commercial farming and agribusiness have developed strongly since 1950; much of the production is exported.
5 Commercialization of agriculture has changed the diet of many Latin Americans, has sometimes cut the supply of staples to the urban areas, and has increased the need for food imports.

4
Migration and urban development

Urban growth

In 1930 most Latin Americans lived in the countryside. Only in Argentina, Chile and Uruguay did a third or more of the population live in cities with more than 20,000 inhabitants (see Table 4.1). Admittedly, there were some large cities; Buenos Aires had more than two million inhabitants, Rio de Janeiro almost one-and-a-half million, and Mexico City one million, but the typical Latin American lived and worked in the countryside. By 1980, however, Latin America had been transformed and almost half of all Latin Americans were living in the cities. Several countries had become as urban as most parts of Europe; in Argentina, Chile and Venezuela two-thirds of inhabitants lived in towns and cities. Only in the poorer countries of the continent were a majority still living in the countryside: Bolivia, the Dominican Republic, Paraguay, and most of Central America.

The urban–rural transition has been achieved through very high urban growth rates. Between 1950 and 1980, the annual growth rate was 4.8 per cent. In places, the pace of urban growth was truly startling; in the Dominican Republic, the urban population doubled every ten years between 1950 and 1980.

Table 4.1 Urban growth in Latin America, 1930–80

	Percentage of total population living in settlements with more than 20,000 persons			Annual growth 1950–80
	1930	1950	1980	
Latin America	17	26	47	4.8
Argentina	38	52	70	2.6
Bolivia	14	20	35	4.3
Brazil	14	21	52	5.7
Chile	32	45	67	3.3
Colombia	10	22	54	5.7
Cuba	26	33	49	2.9
Dominican Rep.	7	12	42	7.1
Ecuador	14	18	40	5.7
Guatemala	11	13	19	4.1
Mexico	14	26	49	5.2
Peru	11	20	49	5.7
Venezuela	14	35	67	5.8

Source: T. W. Merrick (1986) 'Population pressures in Latin America', *Population Bulletin*, vol. 41, p. 23

Migration

Rapid growth rates of this order can be explained only in terms of massive migration from the rural areas. Poor living conditions, combined with high rates of natural increase in the countryside, encouraged many people to move to the cities. We have already seen that land holding in the countryside was very unfair and that there were few efforts at redistributing land until the 1950s. With rural populations rising rapidly after 1940, there was often too little land to support a family. As a result, many young people moved from the farms to the cities; sometimes whole families moved.

The effect of this migration on the cities is easy to demonstrate. In 1964, more than three-quarters of the population of Bogotá aged over 15 years were migrants; in Caracas in 1985, three out of four of the population aged over 45 years of age had been born outside the city.

The huge flow to the cities prompted many to argue that too many people were leaving the countryside. Perhaps the cities would have coped better had the flow of migrants been slower, but it would be erroneous to suggest that most migrants were mistaken in making their move. If we look at rural living standards and compare them to conditions in most cities, migration seems to be a logical decision. There are undoubtedly problems involved in living in the urban areas but at least most families have access to schools, electricity and health care. For the better educated or for those with some skills, the choice of jobs is incomparably greater in the cities.

Unfortunately, many accounts of migration imply that the majority of arrivals are ignorant peasants. While it is true that few are highly literate, the majority do know how to read and write. Indeed, the most consistent finding of migration surveys is that the migrants are typically better educated and trained than the majority who stayed at home. Indeed, it is the selectivity of the migration process that is so consistently demonstrated by migrant surveys. In so far as it is possible to speak of a typical migrant, that person is aged between 15 and 40, has a primary school education, and has some work experience that can be used in the urban environment. This suggests that most rural dwellers do not move out of ignorance but make rational decisions about their urban prospects. They do not move merely because of the 'bright lights' attraction of the city; they move because they have a reasonable chance of succeeding there. The ones who remain in the countryside are those who would be less able to cope in the urban environment: the old, those with unsaleable skills, and those without contacts in the city.

Much has been written about 'push' factors forcing people to move from the countryside. Certainly there is an element of truth in this metaphor, for there are good reasons to leave. The trouble with such an argument is that it does not explain why it is the better-off who tend to move first. If the 'push' factor is so strong, why do the poorest country dwellers not move first? Under conditions of warfare, we would expect all kinds of people to move; if there was hunger, then the poorest would be the most likely to move. As we have seen this is not the case. The very fact that migration is highly selective suggests that other processes are also operating.

Most Latin American migrants know that conditions are better in the city than in the countryside. They are constantly reminded of this by information coming from the city. Since migration has been taking place for a long time, most families have kin in the cities. The city folk tell their rural relatives what life is like in the city and alert them when jobs are available. In any case, most country dwellers have already visited the cities. Buses have connected most parts of Latin America to the urban areas for many years, and fares are not unreasonable. As a result, visits to the city are common, return visits by former migrants even more so; the

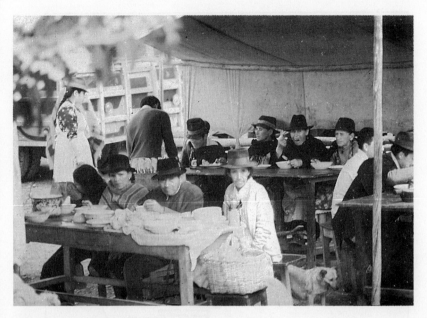

Plate 4.1 Migrants tend to draw from the 15–40 year age group leaving the old behind in the rural areas

result is a regular flow of information between country and city.

In the light of this information families make decisions about whether or not to migrate as a group, or whether it is more sensible for one member of the family to move. Families will often send an older daughter to the city; daughters are perceived to be less useful as farm workers and there is a ready market for domestic servants in the cities. Bright sons and daughters may stay with kin in order to attend secondary school, eventually obtaining better kinds of job. The point is that most of the Latin American poor are highly rational; they cannot afford to make foolish decisions. Migration is not based on myth and emotion; it is founded on informed common sense.

Of course, there are occasions when some people are forced into the cities who are ill-equipped to survive there. Warfare, famine or natural disaster may determine a poor family's destination. At certain times, in particular places, this has been a critical ingredient in migrant decision-making. In Nicaragua, some families have been forced out of border areas by threats from the Contras. In Colombia during the early 1950s, many Liberal Party sympathisers were 'persuaded' to leave their rural

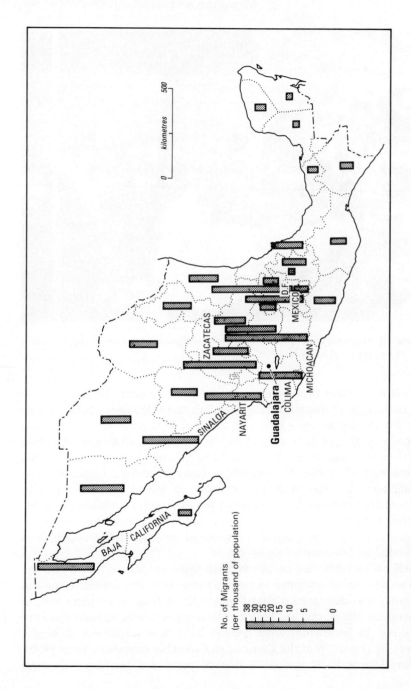

Figure 4.1 Origin of migrants to Guadalajara, 1980

homes; elsewhere, Conservative supporters were similarly forced from the countryside. In north-east Brazil, recurrent droughts have given many desperately poor people little choice but to move to nearby cities. In general, however, this is not the most common explanation for migration. If it were, the migration surveys would not show such a clearly selected population.

One other feature of migration emphasizes that people are highly rational in their decision-making. Rather than travelling long distances to unknown cities, most migrants move relatively short distances to cities they already know. The results of most surveys show that the majority of migrants come from nearby villages and towns. A few have migrated long distances; there are always some adventurers and some who are affluent enough to afford an aeroplane ticket. As Figure 4.1 shows, however, the majority of migrants do not move far. In that figure we record the number of migrants from each Mexican state who are living in the city of Guadalajara compared to the number of people still living in the state of origin. We can see that more than 20 per 1,000 inhabitants have moved to Guadalajara from nearby Colima, Nayarit and Zacatecas, but very few have moved from the distant states of Yucatán or Chiapas. The only clear exception to the short-distance migration rule is the movement from Baja California on the United States border. This exception is explained by the return of people from cities along the United States border, such as Tijuana and Mexicali; many such people have worked illegally across the frontier.

Natural increase versus migration in urban growth

Most census figures show that today's cities contain lower proportions of migrants than the cities of the 1960s. Indeed, as time has gone by a smaller proportion of the population has been born outside the city. Today, most of the young people in the larger Latin American cities have been born there. The reason for this is very simple. Since most migrants arrived when they were between 15 and 40 years of age, they soon gave birth to children in the cities. The rapid migration of young adults led to a rise in the rate of natural increase in the urban areas. Eventually, this raised the proportion of native-born people living in the cities.

Some impression of this change is conveyed in Table 4.2, which shows the proportions for different age groups of people born in Caracas. Among the population aged over 65 in 1981, only 26 per cent were natives of the city; among 25–34 year olds the proportion was 49 per cent;

50 Latin America

Table 4.2 Native-born population as proportion of total population by age cohort, Caracas 1981

Age cohort	Percentage born in Federal District
0–14	86.0
15–24	65.4
25–34	48.8
35–44	35.7
45–54	26.5
55–64	24.7
65+	25.6
Total	62.0

Source: Republic of Venezuela (1985) *Census of Population, 1981*

among those under 15, 86 per cent were natives. While this form of representation slightly exaggerates the point, the tendency for migration to decline relative to natural increase is very clear.

The national settlement system

One of the most remarkable features of Latin American urbanization is the way in which one city dominates the settlement system in so many countries of the region. There are exceptions, notably Brazil, Colombia and Ecuador, but in most nations the largest city is a least four times the size of the second. Table 4.3 shows that *urban primacy* is not a recent phenomenon; indeed, in many countries it had emerged during the colonial period. What is also clear from the table, however, is that the level of primacy has generally increased during the last 50 years. Only during the 1980s have there been signs of a reversal in this trend.

Why did this polarization of population and economic activity occur? Up to 1930 the growth of the major cities was linked to the process of export production. Such cities benefited because they were often ports, were usually national capitals, and always contained a majority of the nation's middle class. Rapid export expansion, therefore, brought a variety of benefits. It sometimes increased export traffic through the port, and usually raised the level of imports. It inflated the government budget, which allowed most appointments to be made to the state bureaucracy. Such an effect was apparent in Argentina where the national capital was located close to the major export-producing area – the Pampas. Export success turned Buenos Aires into a large, cos-

Table 4.3 Urban primacy in selected Latin American countries

Country	Year	Largest city (thousands)	Second city (thousands)	Ratio
Argentina	1923	1,780	266	6.7
	1960	6,739	591	11.4
	1980	9,968	957	10.4
Brazil	1940	1,764	1,326	1.3
	1960	4,574	3,950	1.2
	1980	12,184	8,822	1.4
Chile	1930	713	193	3.7
	1952	1,424	316	4.5
	1983	4,132	565	7.3
Colombia	1938	265	145	1.8
	1964	1,673	948	1.8
	1985	4,208	2,069	2.0
Cuba	1931	654	102	6.4
	1953	1,211	163	7.5
	1983	1,972	353	5.6
Ecuador	1938	160	150	1.1
	1962	511	355	1.4
	1982	1,242	1,049	1.2
Guatemala	1950	347	27	12.9
	1973	979	50	19.6
Mexico	1940	1,803	284	6.3
	1960	5,409	876	6.2
	1980	13,879	2,265	6.1
Peru	1940	614	77	8.0
	1961	1,784	166	10.7
	1983	5,045	539	9.4
Venezuela	1936	263	110	2.4
	1961	1,351	422	3.2
	1981	2,944	889	3.3

Source: Respective national censuses

mopolitan city that dominated most aspects of Argentine life. But similar processes were occurring even in countries such as Chile and Venezuela where the primate capitals were distant from the main sources of mineral exports. In Chile, much of the copper was produced in the far north of the country; in Venezuela, oil was produced in the west and the east of the country. Despite this physical separation, the revenues obtained from copper and oil still found their way into the hands of highly centralized national governments.

More recently, most of the primate cities have increased their dominance as a result both of growing government responsibilities and of the development of manufacturing industry. Since the Second World War,

52 Latin America

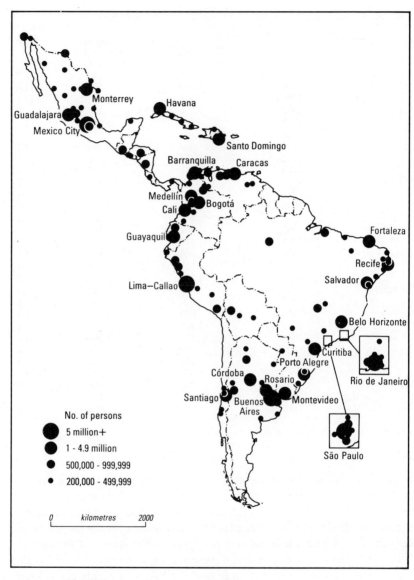

Figure 4.2 Major cities of Latin America, 1985

the process of development has been carefully supervised by national governments. Wages, import licences, prices, electricity provision, trade union activity, labour relations, and many other aspects of industrial life have been monitored and controlled by government. Such centralization has meant that the state bureaucracy has grown. In addition, most of the new industries have established themselves in the primate cities. In part, this has been due to the presence of government; it is easier to negotiate regularly with the authorities if the company is based in the capital. But there were also other obvious attractions, principally that the local market is much larger and more sophisticated than elsewhere, and good communications allow easy contact with the rest of the country. Infrastructure and services are also superior; elsewhere there are regular power failures and a lack of water or telephones. In most countries, industry has located in the largest city – increasing that city's dominance over the rest of the urban system.

In the bigger countries, increasing urban primacy has led to the major cities becoming very large indeed (see Table 4.4). Mexico City's 19 million people make it arguably the world's largest metropolitan area. Buenos Aires has ten million inhabitants, Lima-Callao has around six million. However horrifying these figures may sound, it is necessary to point out that not all large cities are primate cities. In the non-primate urban system of Brazil, for example, São Paulo has grown to around fifteen million people and Rio de Janeiro to nine millions. Even

Table 4.4 Population of Latin America's largest cities

City	Population (thousands)	Year
Mexico City	13,879	1980
São Paulo	12,184	1980
Buenos Aires	9,968	1980
Lima-Callao	5,044	1981
Bogotá	4,208	1985
Santiago	4,132	1983
Caracas	2,944	1981
Belo Horizonte	2,461	1980
Guadalajara	2,265	1980
Recife	2,132	1980
Medellín	2,069	1985
Monterrey	2,001	1980
Havana	1,972	1983

Source: J. Wilkie et al. (1988) Statistical Abstract for Latin America vol. 26, pp. 99–100

Colombia, which has a remarkably balanced urban system, has a national capital which now exceeds four million people. Urban primacy and population size are linked, but not automatically so. This is demonstrated in Paraguay, where Asunción is clearly primate, having six times the population of San Lorenzo, and yet the city has less than half-a-million inhabitants.

Policies to discourage metropolitan growth

Most Latin American governments have expressed concern about the continued growth of their major cities. Such cities are said to be too large, too congested, too polluted and too full of criminals. In so far as these cities are national capitals, their growth is seen as a further symptom of the undesirable effects of political centralization. Decentralization has been widely held to be a desirable goal.

In practice, however, little has been achieved to slow metropolitan growth. Despite a great deal of legislation and the occasional spectacular scheme, neither the rich nor the poor have been convinced that they should live elsewhere. They have stayed in the congested cities because most of the better jobs and the more interesting political, commercial and bureaucratic opportunities remained there. The rhetoric in favour of decentralization has rarely been matched by effective government programmes. Growth centres have been declared in regional plans, but little has been done in practice. New industrial estates have been established in backward regions, but few companies have moved there.

Nonetheless, it would be erroneous to suggest that nothing has been achieved in terms of decentralization. In Brazil, the construction of Brasília during the late 1950s did succeed in removing much of the federal bureaucracy from Rio de Janeiro (see Figure 4.3). In Venezuela, a new steel and aluminium production complex was built at the confluence of the Caroní and Orinoco rivers and a major new city, Ciudad Guayana, constructed to house the workers. In Mexico, a new steelworks led to the construction of Ciudad Lázaro Cárdenas. Recently, plans have been announced to establish a new federal capital in Argentina. Located on the fringe of Patagonia at Viedma, the intention is to move many civil servants away from Buenos Aires.

Some efforts have also been made to divert migrants away from the major cities by creating more opportunities in the poorer regions. Brazil's Amazon programme offers the most spectacular example of agricultural development, but the government has also encouraged industrial dis-

Migration and urban development 55

Figure 4.3 Location of Latin America's new cities

persal by giving large tax incentives to companies moving to Brazil's most impoverished area, the north-east.

The most effective effort to slow metropolitan growth has undoubtedly been in Cuba. After the revolution of 1958, official policy strongly

Case study E

Migration to Bogotá

María Muñoz is 30 and lives in Bogotá, Colombia. She was born in a village 100 kilometres to the north of the city, in the Department of Boyacá. She comes from a large family and has three sisters and four brothers. Because she was the second youngest in the family, and a brother and sister were already working, she was allowed to complete primary school. She can therefore read and write tolerably well. When she was 16 an uncle told the family that one of the managers at the factory where he worked needed a domestic servant. She moved to Bogotá and went to live in the elegant suburb where the manager lived. She was shown how to cook and run the house and stayed with the family for three years. She then met Alvaro, a carpenter who had migrated from a small town 200 kilometres to the north of Bogotá. When they married they moved into rental accommodation half an hour from where she worked, and she continued to help the family on a daily basis, stopping only when she had her first child.

Manuel Contreras is 35 and moved to Bogotá from a small town in the Department of Cundinamarca. The town is only an hour from Bogotá by bus, and he had visited the capital frequently before migrating. He had several friends and a couple of his family living there. Manuel had left school when he was 13 and went to work with his father who was a bus driver. Manuel travelled with his father, helping to collect the fares. Eventually, he learned how to drive and got a job with the same company. When he was 25 years old he had an argument with a supervisor and decided to leave for Bogotá. He went to stay with a friend who also worked on the buses and who helped him get a job. Manuel now drives in Bogotá and hopes eventually to save enough money to buy a taxi. He lived for several years in rental accommodation but when he got married and had his first child the family moved to a self-help settlement in the south of the city. They are still paying off the cost of the plot but because he has regular employment he can afford to employ a skilled labourer to help him with construction of the house. It now has two rooms built of brick and a wooden outhouse with a corrugated roof.

Migration and urban development 57

Case study F

The development of Brasília

The idea of building a new capital in the interior was conceived even before Brazil became independent. It was not until 1956, however, that it became reality. During Juscelino Kubitschek's presidential campaign, he firmly committed himself to this hugely expensive project. Located nearly 1,000 kilometres from Rio de Janeiro, the new site was located in an almost empty area. Through the building of Brasília, a major new impulse would be given to developing the vast undeveloped riches of the Mato Grosso and Amazonia. A further advantage would lie in the removal of federal bureaucrats from Rio de Janeiro, thereby slowing the growth of that city.

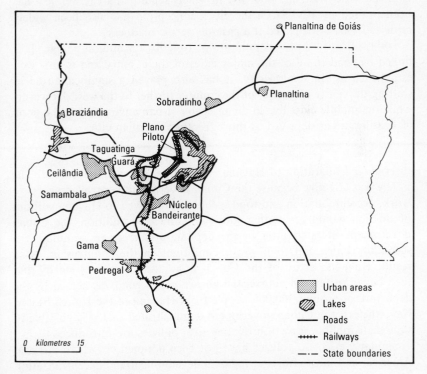

Figure F.1 Map of Brasília – central area and satellite towns

Case study F (continued)

> As the capital of a country with vast potential, the design of Brasília needed to incorporate the latest thinking about architectural and urban planning. In the principal architect's own words, Brasília was designed to be 'simple, logical and beautiful', its development to mark 'a period of optimism, liberty and hope that Brazilians will never forget.' The urban plan incorporated the latest in modernist architectural design, it combined skyscrapers, motorways, open spaces, and futuristic building. Undoubtedly some masterpieces were produced; indeed UNESCO has recently declared it to be part of the 'Cultural and Artistic Heritage of Humanity'. Unfortunately, the plans did not deal satisfactorily with the poor who, in practice, were banished to satellite towns many kilometres from the central city area. Today, Brasília is the most segregated of all Brazilian cities: rich and poor live in widely separated areas. While it has been praised by UNESCO, the city's conception has also been widely criticized; one critic called it a case of 'heroic madness'.
>
> Today, Brasília has a population of roughly 1.8 million people. It is linked by road to all of the major cities of the country and is now well established as the capital city. It has also played a significant role in shifting the centre of gravity of the country further to the west. However, while the middle class live in an agreeable urban environment, the poor of Brasília are as deprived as those in other Brazilian cities.

discouraged migration to Havana, where one in five Cubans lived. The policy was effective because the Cuban government controlled access to work, accommodation and food; people who moved to Havana without permission would get neither jobs nor homes. In addition, the Cuban government invested little in the capital, preferring to stimulate the secondary cities and to improve conditions in the countryside. As a result, Havana's share of the Cuban population remained steady and then began to fall. Of course, the government's goal was helped by the departure of around 300,000 people from Havana to the United States. Nevertheless, the Cuban government demonstrated that it is possible to slow metropolitan growth providing firm policies are introduced.

Similarly effective policies have not been adopted elsewhere in Latin America, and most efforts at regional development and decentralization have been too limited to really slow the pace of metropolitan expansion. Nevertheless, there are now belated signs that the growth of some of the largest cities has begun to slow. There is evidence for Buenos Aires, Caracas, Mexico City and São Paulo of a clear slowing of urban growth; a change that seems to have more to do with growing urban diseconomies, traffic congestion, air pollution, etc., than with government efforts to decentralize. In each case, manufacturing industries have begun to move to towns and cities within reach of, but outside, the metropolitan centre. This trend has been characteristic of the largest cities in the developed world for some years and has now spread to Latin America's foremost metropolitan regions.

Key ideas

1 Most Latin Americans now live in cities, a major change since 1940 when the majority lived in the countryside.
2 Cityward migration was the principal cause of urban expansion but in recent years natural increase has been the dominant source of city growth.
3 Migrants are relatively younger, better educated, and more highly skilled than those who remain in the countryside.
4 Most national settlement systems in Latin America are dominated by 'primate' cities.
5 Most governments have taken action to discourage the continued growth of their major cities, although few have been successful in discouraging such growth.
6 There have recently been signs that some of the largest cities have experienced a measure of spontaneous urban deconcentration; perhaps urban diseconomies now outweigh the economies in those cities.

5
Inside the city

Poor living conditions in the countryside have encouraged many people to move to the cities, because they have relatively little to lose by migrating. The growth of commerce, industry and government in the cities has created new jobs for rural people and offered a means of scraping an income together. Increasingly, however, the cities have grown through their own demographic momentum. The arrival of young migrants changed the age structure of the cities, and the new migrants produced large numbers of city-born children. Today's cities, therefore contain both migrants and natives. Their populations are a mixture of rich and poor, educated and uneducated, skilled and unskilled. However tempted we may be to generalize, there is no typical Latin American city dweller.

The economy of the city

Individual cities vary considerably in their employment structures. Table 5.1 shows that in Ciudad Guayana three workers out of ten are employed in manufacturing. By contrast, only one worker in twenty in Brasília is employed in industry and one person in five works in public administration. In Caracas, one in every eleven employees works in the financial sector compared to only one in seventy in two of the Colombian cities. The different balance of employment in each city is a clear reflection of

Table 5.1 Share of employment in different sectors, by city

	Manufacturing	Finance	Other services
Brasília	5.2	n.a.	83.3
Bogotá	19.8	3.5	62.3
Barranquilla	12.6	1.4	73.3
Medellín	18.4	3.3	63.5
Manizales	15.2	3.8	68.7
Pasto	22.1	1.4	59.0
Ciudad Guayana	29.7	4.0	No data
Caracas	17.3	8.7	No data

Source: Respective national censuses

the different roles they play in the national and international economies. Some cities are national capitals, others ports, others mining towns, others local market centres.

Despite this diversity, it is possible to make some generalizations. First, a much higher proportion of Latin American workers is employed in commerce and services compared with the situation in most developed countries in the past. This is a consequence of the growth of government, the rise of more capital-intensive forms of manufacturing, and, arguably, the sheer growth in the numbers of people seeking work. Second, every city contains a small proportion of affluent people who live comfortably with access to most of the luxuries available to the better-off in developed countries: large homes, expensive cars, country clubs, foreign holidays (see Plate 5.1). These people are employed in managerial and professional activities or receive incomes from rural estates or even foreign investments. They shop in boutiques and department stores and buy their groceries in modern supermarkets (see Plate 5.2). Third, there is a significant proportion of workers who are very poor. Some are children living on the streets, others are old people living in decrepit rental housing on very low incomes, others belong to large, female-headed households where the man has left home. Many of these people get by in the so-called 'informal sector', scavenging at the garbage dump, cleaning cars, or working as domestics or prostitutes (see Plate 5.3).

The formal sector

In most cities the formal sector has been unable to keep up with the increasing demand for jobs. Nevertheless, the growth of manufacturing and white-collar jobs has not been unimpressive. In Venezuela, the

62 Latin America

Plate 5.1 Luxury housing in Rio de Janeiro

number of industrial jobs almost quadrupled between 1950 and 1981, in Mexico a million additional manufacturing jobs were created in the two decades up to 1980, and in Brazil 1.2 million factory jobs were created between 1960 and 1970. Table 5.2 shows that although these are the most exceptional cases, manufacturing employment has also grown impressively in several other countries. Accurate figures on white-collar jobs are more difficult to obtain but in some regions growth in this sector has been even more dramatic. In Venezuela, half a million office jobs were created between 1971 and 1981, an indirect result of the petroleum boom.

Certainly, the number of regular jobs in offices and factories has increased remarkably during recent years. As a result, the formal sector employs a majority of the workers in numerous cities. Such employees are rarely well paid, but they do receive a regular wage, are often unionized, and usually benefit from a variety of social security and welfare payments.

Plate 5.2 Modern department store in Guadalajara

Plate 5.3 Employment at the rubbish dump in Mexico City

Table 5.2 Employment growth in manufacturing industry in selected countries

Argentina	1960	1,885,994
	1970	1,771,250
	1980	1,985,995
Brazil	1960	2,005,775
	1970	3,241,861
Colombia	1964	655,961
	1973	678,322
	1980	1,136,735
Mexico	1960	1,556,091
	1970	2,169,074
	1980	2,575,124
Venezuela	1950	164,685
	1961	284,237
	1971	403,104
	1981	635,185

Sources: J. Wilkie et al. (1988) Statistical Abstract for Latin America, Republic of Venezuela (1985) population census, 1981

The informal sector

Those without formal sector work are employed in the so-called 'informal sector'. The latter includes a wide range of people such as lottery ticket salesmen, bootblacks, scavengers at the local refuse tip, street children, and beggars. Unfortunately, the outer limits of the informal sector are difficult to define, as a huge range of other activity is much less easily categorized. As such, no respectable academic study has ever been able to define precisely what we mean by this beguiling phrase. Perhaps the only clear meaning is that it includes all who are excluded by any specific definition of the 'formal sector' (see also David Drakakis-Smith's book in this series).

Several problems exist in defining the informal sector. First, while a majority of informal workers are poor, there are many who are not. Some of the self-employed are relatively affluent: independent taxi drivers, lottery ticket salesmen with a good clientele, street sellers with a stall at a key junction in the centre of the city, successful drug traffickers. Second, many informal sector workers are paid by formal sector companies. Many seamstresses work at home making clothes to be sold in fashionable boutiques in the best part of town. Some large clothing plants put out work to domestic tailors. Similarly, the collector of paper and cardboard sells it to a large wholesaler or directly to the company that reprocesses it. The lottery ticket salesman often works for that most formal of Latin American enterprises, the government, which uses the profits to pay for public services. If many among the informal sector are actually paid by the formal sector, what does the term 'informal sector worker' really mean?

Living standards

We observed above that some urban dwellers live very well, and many others moderately well. Unfortunately, it is equally true that far too many live very badly. Their housing is crowded and poorly serviced; their diet lacks balance and sometimes even sufficient calories; their illnesses are inadequately treated by deficient health systems. Such problems result from the general economic situation, the unequal distribution of income, and inadequate forms of government intervention. Poverty is a structural condition which is very difficult to resolve. Perhaps the problems are best demonstrated by reference to an example. Let us consider the housing situation in Latin America's cities.

Housing and services

Far too many Latin Americans live in inadequate housing. Clean drinking water is not available in every home and is sometimes lacking in whole neighbourhoods. Levels of crowding are high and the quality of home construction is often very poor. Table 5.3 provides some illustration of the situation in a selection of Latin American cities.

Although housing conditions are all too clearly inadequate, it should not be assumed that all poor people are badly housed. The truth is that there is a great deal of variation. In the self-help settlements, for example, which house somewhere between 30 and 60 per cent of the total population, conditions are highly diverse (see Table 5.4). Many of these settlements have, over a period of ten or twenty years, developed into decent, reasonably serviced neighbourhoods (see Plates 5.4). At the other extreme there are many new settlements where conditions are rudimentary and harsh (see Plate 5.5).

In practice, self-help housing tends to improve with age (see Case study H). What begins as a flimsy hut, perhaps built overnight during a land invasion, often develops gradually into an ordinary, working-class home. Services and infrastructure are eventually provided, either through the

Table 5.3 Housing conditions in selected cities

	% household without:			% households more than two persons/room	% non-owners
	water	electricity	drainge		
Colombia 1985					
Bogotá	4.1	1.6	4.2	29.0	42.9
Medellín	7.8	1.3	8.7	20.0	34.5
Barranquilla	20.5	1.3	26.8	25.0	25.6
Cartagena	29.8	6.1	54.4	30.3	25.5
Mexico 1980					
Guadalajara	8.1	5.0	4.8	36.1	47.7
Tijuana	29.6	10.9	32.6	32.3	48.3
León	16.6	12.4	22.6	48.1	37.3
Venezuela 1981					
Caracas	8.9	6.3	8.2	n.a.	36.8
Maracaibo	16.5	5.8	16.9	n.a.	20.2
Mérida	7.0	5.5	8.5	n.a.	40.8

Source: Respective national censuses

poor stealing water and electricity or because the authorities begin to supply the services. The process of housing improvement takes far too long but improvement is widespread and continuous.

Table 5.4 Relative growth of irregular settlement in selected cities

	Year	City population (thousands)	Population in irregular settlements	Percentage
Caracas	1961	1,330	280	21
	1964	1,590	556	35
	1971	2,220	867	39
	1985	n.a.	n.a.	61
Lima	1961	1,846	317	17
	1972	3,303	805	24
	1981	4,601	1,150	25
	1981	4,601	1,460	33
Buenos Aires	1956	6,054	104	2
	1970	8,353	434	5
	1980	9,766	957	10
Rio de Janeiro	1947	2,050	400	20
	1957	2,940	650	22
	1961	3,326	900	27
	1970	4,252	1,276	30

Source: Respective national censuses

Inside the city 67

Plate 5.4 Self-help housing can improve through time – Bogotá

Plate 5.5 Flimsy housing – a land invasion on the fringes of Bogotá

Plate 5.6 Rental housing: a *vecindad* in a consolidated self-help settlement in Guadalajara

Even if self-help housing tends to improve with age, we should not assume that all who live in such accommodation actually own it. In some cities of Latin America as many as half the population rent accommodation or share with kin. Indeed, the self-help settlements contain many houses·where the owners rent out rooms to tenant families. The more consolidated and older parts of a city are full of buildings accommodating large numbers of tenant families. Plate 5.6 shows the kind of rental housing that still exists in parts of Latin America.

There is a widespread assumption that building a self-help home in a Latin American city costs nothing. Such an assumption is valid only if land can be invaded, if flimsy materials are used in construction, if such materials can be obtained for nothing, and if electricity and water can be stolen from nearby supply lines. There are very few cases, however, where all of those conditions are met.

For a start most governments, and certainly most landowners, oppose invasions. In almost all cities, plots of land have to be purchased. The typical self-help dweller in urban Brazil, Colombia or Mexico has purchased a plot in an illegal settlement. The settlement is illegal because it lacks the services necessary to secure planning permission. The

Plate 5.7 Unsuitable land is often used for housing construction – a house with a view in Tijuana

settlements may also be located on land unsuitable for urban development, the slopes being too steep or the land liable to flooding (see Plate 5.7). Despite these conditions, a price still has to be paid to the subdivider of the land. Such a price is often high when compared to typical wage levels and it will take several years to accumulate the necessary savings. As a result, subdividers sometimes sell on a loan basis, a 10 per cent deposit followed by the rest paid over a period of four years.

If land has to be purchased, so also do the materials. In cities as large as São Paulo or Mexico City, millions of people are busy constructing their own homes. While some corrugated iron, bricks, wood and cardboard is no doubt lying around on the streets, someone else will usually have claimed it first. As a result, most poor people have to purchase their materials; they buy bricks, doors, water tanks and window-frames from local building merchants.

The installation of services may also cost money. Some water and electricity is stolen, but this is not possible if there are no service lines nearby. Under such circumstances, the service has to be purchased. Water is sold by private companies who send tankers round to fill the oil drums that are used to store it (see Plate 5.8). Alternatively, public

Figure 5.8 A water tanker filling the oil drums in a self-help settlement

agencies supply the settlements at cost.

There is no doubt that self-help housing does provide many of the poor with an adequate home. But it is only the less poor who can afford to buy the necessary land, materials and services; the really poor are often excluded from the self-help process. Such families are forced to rent rooms, or to share accommodation with kin. Given the severity of the 1980s recession, the fear must be that more and more people are being forced to live in rental and shared accommodation. Do-it-yourself works best when people have the money to pay for materials and services.

Political activity and protest

Urban protest in Latin America is common. Poor people, organized by unions or neighbourhood groups, protest in the street about their living standards or about some failure of the government. Most days, in the main square of Mexico City, there are marches organized by tenant associations, opposition parties or trades unions. During the recent military regime in Argentina there was a regular gathering of women in the *Plaza de Mayo* in Buenos Aires, mourning for and protesting at the disappearance of their children in the so-called 'dirty war'. In numerous Colombian towns there have been civic strikes to protest about deficient services and the lack of infrastructure. In Bolivia, newly redundant miners marched to protest at the closing of the nationalized tin mines. Recently in Caracas there were riots and raids on supermarkets in protest at the rising cost of living. Throughout the region, rises in the cost of transportation have stimulated protest, not infrequently culminating in the blocking of roads and the burning of buses.

That the poor should protest in Latin American cities seems entirely understandable. Many factory workers, most of whom have never been very well paid, have seen their incomes fall dramatically during the 1980s. The majority of the poor are housed inadequately and lack satisfactory public services and infrastructure. Transport facilities are slow, congested and uncomfortable; health services for the poor are generally unsatisfactory. And yet, given these problems and the marked inequalities in wealth found in the cities, it is surprising how little the poor protest. It is especially surprising since so many Latin American governments have lived in fear of the revolutionary potential of the urban poor. The US government continues to advise its Latin American counterparts about security; the fear that the shanty towns will turn 'red' remains, in some circles, a constant concern.

In these circumstances, why is there not more overt political hostility towards the state? Why *do* the poor not protest more? The answer seems to lie in four broad features of Latin American society. First, urban protest requires organization, and many political groups have found negotiation with the government to be a sounder policy than confrontation. Through a combination of carrots and, frequently, big sticks, Latin American governments have discouraged unions from striking and most neighbourhoods from protesting. Often, it is dangerous to protest too vociferously. Leaders may be arrested, demonstrators may be attacked by the police, invaders of land may have their shanty homes demolished. In addition, trouble-makers may be excluded from the list of potential beneficiaries of jobs, services, grants or pay rises.

Second, the chances of confrontation have been reduced by the state co-opting the leaders of the poor. If effective leaders appear in low-income areas or in factories, and begin to organize some kind of protest, then many governments have found it best to co-opt those leaders. They may offer them jobs in the government bureaucracy, give them political posts, or offer a retainer to supplement their current incomes, all in

Case study G

Earning a living in the informal sector

Juan Alvarez is 35 years old. He is a bootblack in Guadalajara, cleaning shoes every day except Sunday. In order to work in one of the central squares he had to buy a licence from the bootblack who had previously occupied his site. Without the licence he would have been constantly harassed by the police and by other bootblacks. The licence was not cheap, it cost the equivalent of three months' takings, but he had saved the money from his previous job in a small local factory. He had worked in the factory for five years, but found the job hard and the working conditions claustrophobic. When his wages stopped rising in line with prices, he decided to leave. While he would earn no more as a bootblack at least he could work outside, a pleasure in the agreeable climate of Guadalajara. He began to supplement his income with the sale of lottery tickets. At first, sales were a bit irregular but as he has built up his shoe-shining clientele things have improved (see Plate G.1).

María Alvarez also works in the informal sector. Unlike Juan she

Plate G.1 The informal sector – bootblacks at work in central Guadalajara

Case study G (continued)

works from their home, running a shop in a self-help settlement in the east of the city. She started by selling soft drinks and sweets, but now stocks a range of groceries as well. She is not really sure how much she makes from the store because she does not keep proper accounts, and because food for the family's own meals is taken from the shelves. Business is certainly not very good – a fairly affluent merchant recently set up a store nearby and there is also competition from a supermarket fifteen minutes' walk away. The main advantage of running the store from home, however, is that she can earn some money while looking after her own, and the neighbour's children. The neighbour gives her money over and above the cost of the food that the children eat during the day.

return for dampening rather than stimulating protest. This kind of action is part and parcel of the normal political activity of Latin America. It is in this way that the potentially troublesome, such as the petroleum workers of Mexico, have been controlled.

Third, protest has been limited because most poor Latin Americans are so busy trying to earn enough money to survive that they have relatively little time for political action. Working hours are often long, and so too is the journey to work. Frequently, both husband and wife will be earning during the day and will be constructing their homes during their 'spare' time. Add the intensive demands made by the extended family, an important feature of Latin American society, and there is relatively little time left over for protest. Only under the direst of circumstances will wrath outweigh the pressure of time. In any case, the state has been successful in convincing most Latin Americans that protest is antisocial. The self-help settlements contain more political conservatives than radicals. The ideology that the state is trying to help the poor has been sufficiently well diffused to help keep many people off the streets.

Fourth, although social conditions are poor, for many years there were clear signs that they were improving. Until the current crisis, self-help settlements did gradually receive water and light. Buses did reach the settlements and councillors did deliver funds with which to build a community hall or improve the football pitch. In an environment where most settlements are receiving some help and resources can be found to bring about marginal improvements, petitions seem to be more appropriate than protest.

Case study H

Building a self-help home

Alfonso and Isabel Rodríguez live with their three children in a low-income settlement in the south of Valencia, a Venezuelan city with around 800,000 inhabitants. The population has grown rapidly as manufacturing plants have moved to the city in preference to the crowded conditions found in Caracas. Alfonso works in the Ford plant. He was born in the mountain state of Mérida and arrived in Valencia in 1975 after an uncle told him that work was available. He stayed with his uncle for a while in a consolidated self-help house in the south of the city and then moved into rental accommodation nearer the centre. He met Isabel, who had been born in Valencia, six months later. They rented a new home together and stayed there for eighteen months. Two years later he managed to obtain a plot of land through an invasion. The invasion was organized by an employee of the local authority who was trying to win support for a councillor who was hoping to be re-elected that year. One hundred and twenty families established rudimentary shacks early one Sunday morning. Because of the protection given by the councillor there was little trouble from the police and the settlers soon began to improve the accommodation. Since Alfonso had a reasonably well-paid job, he could afford to buy cement and bricks with which to improve the house. In addition, he sold half of the 20 by 35 metre plot he had obtained to two other families; this was sufficient to buy the rest of the materials he needed. He could also employ, on an occasional basis, a friend who worked in the construction industry. Progress on the house was slow but steady. Since he had to work at his paid job during the week, he could do little except at weekends. Even then there were interruptions, family visits, the occasional fiesta, demands by the children to go out. Isabel helped with some of the lighter jobs when she could, but three rapid pregnancies and a miscarriage limited her participation in the actual building.

By 1985, five years later, the house had three rooms. It was not pretty to look at but it was solidly built and the roof kept out the rain. It had electricity, stolen from the mains by a neighbour who worked for the electricity board. Water has been provided by the government during the last election campaign, but sewerage was still lacking. There was a school in the next settlement, and the eldest child would start to attend it in a couple of years' time.

Inside the city   75

Key ideas

1 The size of the formal sector has grown dramatically in most Latin American cities, but has been unable to absorb all the new workers demanding jobs.
2 Many Latin Americans work in the informal sector because they cannot get formal sector work.
3 The informal sector is highly diverse and includes both the affluent and the very poor. Many workers are employed indirectly by the formal sector.
4 Self-help housing accommodates a majority of the population, but many families rent rooms from the owners.
5 Self-help housing improves through time but the rate of improvement depends upon the cost of land, materials and services.
6 As the recession of the 1980s has cut real incomes, consolidation of the self-help home has become more difficult.
7 Political protest is not uncommon in Latin American cities, but given the general living conditions it is perhaps surprising that it does not break out more frequently.

6
Poverty, development and inequality

Social conditions

There can be little doubt that living conditions for most Latin Americans improved gradually after 1930 (see Table 6.1). Life expectancy increased dramatically, health standards improved, and levels of literacy rose markedly (see Plate 6.1). Although the quality of life began to improve much earlier in the south of the continent, even in Central America and

Table 6.1 Social conditions by country, 1950–85

	Illiteracy		Life expectancy		Urban pop. with piped water		Calorific intake	
	1950	1985	1950–55	1980–85	1960	1980	1965	1981–83
Argentina	13.6	4.5	62.7	69.7	65.3	67.9	2,868	3,195
Bolivia	67.9	25.8	40.4	50.7	55.8	24.1	1,731	2,061
Brazil	50.5	22.3	51.0	63.4	54.7	50.7	2,541	2,564
Chile	19.8	5.6	53.8	71.0	73.8	94.8	2,523	2,662
Colombia	37.7	17.7	50.7	63.6	78.8	69.0	2,220	2,543
Ecuador	44.3	17.6	48.4	64.3	58.2	78.7	1,848	2,052
Guatemala	70.7	45.0	42.1	59.0	42.3	51.3	1,952	2,189
Mexico	43.2	9.7	50.8	67.4	67.5	61.8	2,623	2,966
Peru	38.9*	15.2	43.9	58.6	47.3	60.8	2,255	2,150
Venezuela	50.5	13.1	55.2	69.0	54.5	89.1	2,392	2,664

Sources: J. Wilkie et al. (1988) Statistical Abstract for Latin America, vol. 26
*1960

Plate 6.1 Educational provision has improved, even in the rural areas

the Andes life eventually got better. In addition, many Latin Americans began to gain access to some of the benefits of a consumer society. In Brazil in 1976, for example, one in six families owned a car, three out of four had a radio, two out of five a refrigerator, and almost half had a television set.

Unfortunately, the general improvement was rudely interrupted by the severe economic recession of the 1980s. The recession had led to a marked increase in poverty; in several countries, living standards have fallen to the levels characteristic of ten or fifteen years earlier. Between 1981 and 1989, per capita gross domestic product fell 27 per cent in Bolivia, 25 per cent in Venezuela, 18 per cent in Guatemala, 24 per cent in Argentina and 25 per cent in Peru; it increased in only four countries: Cuba, Colombia, Chile and the Dominican Republic. Between 1980 and 1989, the effects of inflation cut the income of private sector manual workers in Lima by 70 per cent, and in Mexico manufacturing workers saw their wages fall by 30 per cent. Rates of open unemployment rose in most cities (see Table 6.2). There were also unmistakable signs of rising levels of malnutrition and infant mortality in several Latin American cities.

Table 6.2 Open unemployment rates in selected cities

	1980	1981	1982	1983	1984	1985	1986	1987	1988
Buenos Aires	2.3	4.5	4.7	4.2	3.8	5.3	4.6	5.4	n.a.
Bogotá	6.8	5.2	6.7	8.7	11.7	12.6	12.6	10.5	9.4
Medellín	14.7	13.1	15.2	16.2	16.9	15.4	14.5	11.3	11.7
Mexico City	4.3	3.9	4.0	6.3	5.8	4.9	5.1	4.5	n.a.
Lima	7.1	6.8	6.6	9.0	8.9	10.1	5.4	4.8	7.9
Montevideo	7.4	6.7	11.9	15.5	14.0	13.1	10.7	9.3	9.2
Santiago	11.8	9.0	20.0	18.9	18.5	17.2	13.1	11.9	11.2
Rio de Janeiro	7.5	8.6	6.6	6.2	6.8	4.9	n.a.	n.a.	n.a.

Sources: Colombia (DANE) *Boletín de Estadística*, various years. United Nations, *Economic Survey of Latin America*, 1982 and 1985. Economic Commission for Latin America (1989) *Preliminary Overview of the Latin American Economy 1989*

The recession has hit the urban poor particularly hard. The rich were able to move their money into US dollars, thereby avoiding the worst consequences of domestic inflation. The poor have suffered greatly from the combination of high rates of inflation and limited pay rises. Acute social differences, which have long been a feature of Latin American society, have become more marked.

Race and ethnicity

Race and ethnicity are important components in social differentiation in Latin America. Although the divide between indian and white, or white and black, is never total, there are clear signs of racial discrimination in most parts of the continent. Ask most white Latin Americans about indians or blacks and a joke will usually be the response. If one looks at the political élite of most countries, it is noticeable that practically every president has been drawn from the 'European' sector of the population. Of course, there has been the odd exception over the years, notably in Mexico where an indian, Benito Juárez, became president in 1858. In general, however, indians and negroes do not attain positions of political or economic influence in Latin America.

The fact that there is little upward mobility does not mean that there is constant racial tension. Brazil in particular prides itself on its lack of racial conflict, and there are numerous cases of famous black footballers and entertainers who have made a lot of money from their trades. Since 'money lightens the skin' in Brazil, there is a certain degree of social mobility. Nonetheless, having a black skin is still a social disadvantage, since everyone's colour is graded along a continuum ranging from whiteness to blackness. Having a darker skin is undoubtedly a barrier to entry into the higher echelons of society.

Elsewhere in the region, the barriers to upward mobility for indians and negroes are probably still greater. Ethnic typecasting, which is very common in most countries, is one such barrier; indians, for example, are often assumed to be illiterate. However, typecasting is also closely tied to class, and judgements are based as much on dress and occupation as on skin colour. People who work in the countryside may be referred to as indians, even when they are light-skinned.

Language, too, forms a major barrier for the substantial minorities who speak indigenous languages in Bolivia, Central America, Ecuador, Mexico and Peru. It is much harder to reach the top of the political or economic tree in a language other than one's mother tongue. Of course, as society becomes more urbanized, language poses fewer problems because the majority of the population begin to speak Spanish or Portuguese. However, monolingualism does not overcome social discrimination, and new forms of differentiation are generated.

Social class

Portuguese and Spanish rule introduced a particular form of class segmentation into Latin America. While race was an important element in determining economic and social status, it was the ownership of land that was the key to class position. The élite controlled land, and most of the rest of the population were expected to work that land. Today, ownership of land no longer has quite the same social cachet. Ownership of a large estate is clearly a significant status symbol but there are now other routes to high social position: manufacturing, commerce, the media, transport, urban real estate and so on. There is also a further route towards power and influence, i.e. through the state bureaucracy. The routes to the top can be through political office, through bureaucratic performance, or through a military career. Class does influence who reaches the top, but some from the lower orders do make it. While there are considerable differences between countries – it is easier for the poor to reach the top in Argentina or Cuba than in Colombia or Mexico – few countries still possess a rigid 'caste' system.

In other respects, however, Latin America has changed very little. The peasantry and landless rural workers remain at the bottom of the social order. With few exceptions the peasantry control little in the way of land and even less in terms of resources. Only on rare occasions have the

Table 6.3 Secondary and university enrolment in selected countries, 1960–80

Country	Percentage of students enrolled as share of eligible age group			
	12–17 years		18–23 years	
	1960	1980	1960	1980
Argentina	48	73	13	37
Bolivia	29	54	5	17
Brazil	30	59	5	32
Chile	55	87	7	22
Colombia	29	64	4	33
Cuba	43	83	7	30
Mexico	37	67	5	18
Peru	43	84	13	33
Venezuela	49	61	9	24

Source: J. Wilkie *et al.* (1988) *Statistical Abstract for Latin America*, vol. 26, p. 158

landless and landed peasantry wielded political power, and then usually as a result of armed insurrection. Normally they have been easily controlled by the dominant groups.

Large numbers of rural people have moved to the cities and joined the urban working classes. Such a vast group hardly constitutes a single class for it includes a unionized industrial workforce as well as the vast and heterogeneous ranks of the informal sector. The 'class' is divided in many ways. The workers in mining, power supplies, ports and railways are able to demand better pay settlements and working conditions than many other groups. This sometimes makes them more militant than other workers, and it certainly does little to cement working-class solidarity. On the whole, such organized groups gain access to social security schemes and obtain special benefits, leaving most of the urban poor behind. The informal sector, by its very nature, is difficult to organize and is divided in a multitude of ways. As a result, it is easy for governments to manipulate the differences between working-class groups and, in the process, to suppress any opposition.

Between the élite and urban working classes there is an increasingly large middle sector. As economic development has progressed, many new jobs have been created in professional and clerical activities. As education facilities have improved, more and more people have attended secondary school and even university (see Table 6.3). This middle sector now includes one-fifth of the urban population in the more prosperous cities. It has gained access to many of the benefits created by decades of economic growth – cars and home ownership, education for the children, health care and holidays.

Gender

An additional division in Latin American society is created by the gender roles imposed on men and women. *Machismo*, the belief that males are superior to women, gives rise to stereotypical behaviour – a belief in male fearlessness, sexual prowess and control over women. In contrast to the idealized behaviour of the male, the 'correct' attitude of the female is to be submissive, kind, caring and gentle. Different labour roles are linked to these gender stereotypes – the man is seen to be the breadwinner, the woman is the home-maker and the one responsible for maintaining harmony in the home.

The consequence of these kinds of role models is that many men, particularly among the *mestizos*, control most of the best jobs and keep an excessive proportion of their wages for themselves. The woman is given the difficult task of running the home on a very limited income. Relatively few Latin American women outside Cuba go out to work; indeed women make up a lower proportion of the Latin American labour force than in any other major world region (see Table 6.4). Even when women choose, or are forced, to go out to work, only a limited range of low-income activities are open to them. Domestic service, a job in retailing, or at best secretarial work, are the main kinds of outlet open. While factory jobs are available in places, pay rates are still lower than those for men. For educated middle-class women the range of options has been increasing in recent years but, for the majority of Latin American women, the availability of decently paid work is very limited. This is particularly serious given that a considerable proportion of low-

Table 6.4 Female participation in the labour force by major world region, 1975

	Economically active women as proportion of total female population (%)
Worldwide	28
Latin America	12
Middle East	22
Africa	23
North America	28
Europe	29
Far East	36
USSR	48

Source: Adapted from Table 4.1 in T. Cubitt (1988) *Latin American Society*, p. 106

income households now have female heads, many women having been deserted by husbands or male companions.

The distribution of income

In 1976 the average income of the richest 5 per cent of Brazilians was 58 times that received by those in the poorest quintile. Brazil is hardly an exception in Latin America, for income and wealth is very unfairly distributed in most countries. Table 6.5 shows that the poorest one-fifth of the households of seven Latin American countries receive a very small share of total household income. While the table also shows that the poor are not well rewarded in Britain, the United States or Japan, in these countries they receive a more generous share than equivalent groups in Latin America. The contrast with the three developed countries is still more marked when the income of the most affluent 10 per cent of households is considered. Whereas the top 10 per cent of the households in the three developed countries receive about 23 per cent of total household income, in Latin America this group of the population receives somewhere between 30 and 51 per cent of the total.

It was long assumed that the growth of per capita income would automatically lead to greater equality in the distribution of income. Historical studies showed how industrialization in most of the developed countries was accompanied by a gradual reduction in inequality. In Latin

Table 6.5 Inequality of income in Latin America

Country	Year	Percentage share of household income by household group		
		Poorest 20%	*Poorest 40%*	*Richest 10%*
Argentina	1970	4.4	14.1	35.2
Brazil	1972	2.0	7.0	50.6
Costa Rica	1971	3.3	11.8	39.5
El Salvador	1967–77	5.5	15.5	29.5
Mexico	1977	2.9	9.9	40.6
Peru	1972	1.9	7.0	42.9
Venezuela	1970	3.0	10.3	35.7
USA	1980	5.3	17.2	23.3
UK	1979	7.0	18.5	23.4
Japan	1979	8.7	21.9	22.4

Source: World Bank (1987) *World Development Report*, pp. 252–3

Poverty, development and inequality   83

Table 6.6 Real incomes by socio-economic group in Brazil, 1960–76 (1970 US dollars)

Quintile	1960	1976	% increase
Bottom fifth	59	75	27.1
20–40%	135	183	35.6
Middle fifth	234	309	32.1
60–80%	342	607	77.5
Top fifth	916	2317	152.9
Top five per cent	1869	5475	192.9

Source: C. Brundenius (1981) 'Growth with equity: the Cuban experience (1959–80)', World Development, vol. 9, p. 1090

America, however, there seems little sign that the distribution of income is improving through time.

In Brazil there is even evidence that a deterioration occurred between 1960 and 1976. The poorest 40 per cent of the population saw their share of income decline from 10 per cent to only 8 per cent; the richest 5 per cent increased their share from 35 to 39 per cent. Admittedly the poor did not get absolutely poorer, but the benefits they received were infinitessimal compared with those received by the top 5 per cent of the population (see Table 6.6).

In Mexico between 1950 and 1977, the distribution of income changed little in general (Cordera and Tello 1984: 268). In one respect, however, there was a tendency to greater equality, for the richest 5 per cent of the population saw their share of income decline from 35 per cent to 25 per cent. Unfortunately, the redistribution did not benefit the poor because

Table 6.7 Distribution of income in Cuba, 1953–73

Quintile	1953	1973
Bottom fifth	2.1	7.9
20–40%	4.1	12.5
Middle fifth	11.0	19.2
60–80%	22.8	25.5
Top fifth	60.0	34.9

Source: Adapted from C. Brundenius (1981) 'Growth and equity: the Cuban experience (1959–1980)', World Development, vol. 9, p. 1090

Case study I

The socialist model in Cuba

Cuba was a Spanish colony until 1898 when it was invaded by the United States. It gained political independence in 1902 but immediately became very closely tied to the United States' economy. During the next half-century Cuba produced sugar for the United States, provided North Americans with cheap leisure facilities, and acted as a market for US manufactures. Cuba was administered by a series of corrupt and incompetent regimes. This period was ended by a revolution led by Fidel Castro in 1959, who established a Marxist regime on the island.

In 1959 a major land reform expropriated the large foreign-owned estates and placed most of the land under government ownership. The rest was redistributed among former tenants and smallholders. A major attempt was made to improve rural conditions, a literacy campaign and a major health improvement programme proving particularly successful.

Economic strategy has changed course several times but the most consistent feature has been continued dependence on export production combined with a major effort to industrialize. At first, economic growth rates were very disappointing but after 1970, and the introduction of a Soviet-inspired planning system, rising productivity resulted in impressive rates of economic growth.

Table I.1 Economic growth in Cuba, 1962–88

	Gross social product	Gross social product per capita
1962–65	3.7	1.3
1966–70	0.4	-1.3
1971–75	7.5	5.7
1976–80	4.0	3.1
1981–85	7.3	6.4
1986–88	-0.7	-3.5

Source: 1962–85: A. Zimbalist and S. Eckstein (1987) 'Patterns of Cuban development: the first twenty-five years', *World Development* vol. 15, p. 8. 1986–88: United Nations Economic Commission for Latin America (1988) *Preliminary overview of the Latin American Economy*, 1988

Case study I (*continued*)

Today, Cuba is the world's second largest sugar producer and accounts for around one-quarter of the world's sugar exports. That commodity still constitutes the major export of the country, making up 77 per cent of total exports in 1982. Ever since the revolution most of the sugar has gone to the Soviet Union, the latter paying consistently more than the international price. In 1986, for example, when the international price was four cents per pound, the USSR was thought to be paying 40 cents. The close relationship with the USSR arose because the United States has maintained an economic embargo against the island ever since the revolution. The Soviet Union took advantage of this conflict to gain a major ally within 150 kilometres of the United States' coast.

Cuba's immediate economic future, however, looks uncertain. Its problems arise from being too dependent on sugar, its over-reliance on the Soviet Union, its considerable foreign debt, a deteriorating balance of payments, and from worryingly low levels of economic efficiency.

If there are signs of problems on the economic front, Cuba's social record has long been a beacon of progress compared to those of most other poor countries. In 1978, 1.1 million Cubans attended secondary school compared to only 90,000 in 1958; today, literacy rates are very high even in the rural areas. Major improvements in health care and in general living conditions have reduced death rates. Life expectancy, already high before the revolution, has improved from around 60 years in 1958 to 74 years in 1984. Fortunately, the fertility rate has also fallen: in 1973, it was 122 per 1,000 women aged 15–44 years, by 1981 it was down to 60.

Employment on the island is guaranteed, although official reports still recognized a rate of 3 to 4 per cent unemployment in 1981. Labour participation rates are high, and some 38 per cent of women were employed in 1984, a very high figure by Latin American standards (see Table 6.4). If there is an employment problem on the island, it is that most Cubans work too many hours, the result of highly inefficient work practices.

Cuba has been supported by the Soviet Union, and opposed by the United States, its traditional trading partner. Both relationships are critical in understanding Cuba's recent development: the social record could not have been maintained without Russian support, as the US embargo undoubtedly damaged the Cuban economy. Today, Cuba's

Case study I (continued)

dependence on the Soviet Union matches its previous dependence on the United States. As Turits (1987: 77) points out:

> In the last instance, the dynamics of the Cuban economy are conditioned by political decisions made in Moscow over subsidies and trade patterns. . . . Yet evidence is scarce that current dependence on the Soviet Union shapes and delimits Cuban society in a manner analogous to Cuba's former dependence on the US. No one contends that Cuba has been subject to long-term capital outflows, deteriorating terms of trade, or extraction by the 'core'. And dependence on the Soviets entails neither foreign ownership of the means of production, nor, apparently, any intimate nexus with the economic vicissitudes of the socialist 'center'. Further, those socioeconomic indices normally associated with dependency – uneven development of town and country, and increased economic inequality – are conspicuously absent, indeed reversed, in Cuba today.

Overall, therefore, the Soviet Union's influence seems to have been rather benign. As a result, most Cubans have gained considerably from the revolution.

However, it is rather doubtful whether the Soviets will support other socialist countries in the region quite as generously. Certainly, their help to Allende's Chile was rather limited, and Nicaragua has received much less help than Cuba. In this sense, it is rather hard to judge the value of the Cuban experience for other countries in the region.

---

the poorest 40 per cent of the population saw their share decline from 13 per cent to 11 per cent. The main beneficiaries of change were the middle classes.

It is only in Cuba that there has been a serious effort to remedy the inequality of income (see Table 6.7).

The development model

The majority of Latin American countries have followed some kind of capitalist development path. Only in Cuba since 1959 (see Case study I), and for brief periods in Chile, Nicaragua, and Peru, has there been any

Plate 6.2 Modern values have diffused rapidly through Latin America, often mixing with more traditional beliefs: a roadside shrine

significant move toward a socialist model. The general strategy for development has not been significantly different from that followed in the United States and Western Europe. It has relied on economic growth to create the resources for redistribution to the majority of the population. The model has worked less well than in the developed countries partly because of fast population growth and partly because economic growth has been less rapid than had been hoped for. The poor benefited to some extent, at least until the 1980s, but less than most other social groups. Certainly, there have been few efforts at redistributing income through land reform, progressive taxation, and controls on urban land speculation.

The capitalist development model has relied heavily on industrialization to create jobs. Industry has usually been protected by high import taxes and by exchange-rate policy. Only relatively recently have greater incentives been given to exporters and a more determined attempt made to expose domestic enterprise to external competition. However, at its most effective, as in Brazil between 1967 and 1974, the industrialization model was successful in achieving rapid economic growth (see Case study J).

Case study J

The Brazilian economic miracle

During the early 1960s Brazil began to face serious economic and social problems. Economic growth had slowed down, inflation had risen dramatically, and there was a difficult political situation. This series of problems convinced the military that they should take over the country; they were to remain in power until 1985.

The military government introduced what is known as a 'bureaucratic–authoritarian' model of development. It was called 'bureaucratic' because most of the key decisions were made by a technically trained state élite. While the private sector was a strong beneficiary of the development process, the state performed an impressive range of economic and social functions. The model was 'authoritarian' because it allowed little political debate and nothing in the way of social complaint. Trade unions were rendered impotent, rural demands for land reform were ignored, and the opposition was imprisoned or otherwise silenced.

The new economic strategy aimed to slow inflation and then to encourage growth through increasing export production, particularly of manufactures. Inflation was slowed through strong controls over wage rises, and trades unions were not permitted to protest. Brazil's exports were made more competitive and the government also provided companies with substantial export subsidies. In response, manufacturing industry was successful in increasing its share of the country's exports; by 1979, manufactures made up 11 per cent of total export revenues.

For a while, the economy boomed. Between 1967 and 1974 gross domestic product was growing at 10 per cent annually. Even after the shock of the OPEC petroleum price rises, economic growth continued. The pace of growth was really quite dramatic. In 1965 the country produced 185,000 road vehicles, by 1972 annual production was up to 611,000 and by 1980 had risen to 1.2 million units. Steel production rose from 4.4 million tonnes in 1968 to 12.3 million tonnes in 1978.

Of course, the distribution of the benefits from growth was highly unequal (see Table 6.6). At the same time, the poor did benefit through higher incomes, and a larger proportion of the urban population had access to consumer goods. By contrast, social improvements were limited and, despite their promises, the military did little to improve education or health care.

Case study J (continued)

The basis of economic success was also under threat. Increasing levels of debt and the beginning of the world recession put the export model under increasing pressure. The cost of servicing the debt rose inexorably – from 12 per cent of exports in 1970 to 85 per cent in 1985. By the late 1970s there were clear signs of the inefficient use to which many foreign loans had been put, and more and more examples of corruption were coming to light. In recognition of the growing problems, the military began to discuss the possibility of a return to democratic rule. They allowed partial political representation and eventually an elected president returned in 1985. The military left what they had inherited, a severe economic crisis.

Despite the widespread adoption of a capitalist development model, most of the more successful countries have a large and usually expanding public sector. The state's role was both economic, in the sense of establishing basic industries (such as steel) and building infrastructure (such as roads, ports and power facilities), and social, in the sense of improving education, health care and housing. This combination of roles had led to a very powerful state apparatus developing in several countries. Indeed, in Argentina, Brazil and Mexico, state participation in the gross domestic product was around 30 per cent in the late 1970s. Only recently have government efforts to encourage privatization begun to reduce the level of state involvement. In the last few years the Mexicans have closed or sold off numerous state enterprises including the main airline, a major steelworks, petrochemical plants, most of the banks, and a hotel chain; the number of state enterprises has fallen from just over 400 in 1982 to just over 100 in February 1988. Similarly, determined efforts at privatization have also been characteristic of Argentina, Bolivia and most of all Chile. In this respect, Latin American governments have been following the trend set in many developed countries such as Britain and Japan.

The nature of the state

Since Latin American independence, the military has interfered regularly in national policies. Most countries have experienced long periods of military command; very few have had extensive experience of civilian rule. The lack of true democracy and the tradition of political instability is not, however, peculiar to Latin America; many non-Latin countries have suffered either from regular coups and countercoups or from continuous changes of government. Indeed, political instability seems almost to be the normal pattern in most parts of Africa and Asia, and it is only quite recently that mature countries, such as France, Greece, Italy and Spain, have managed to establish relatively stable systems of democratic government. The truth is that it is very hard to establish and maintain democracy in countries with acute economic problems, unequal access to power and little tradition of democratic rule.

As a result, what has most characterized Latin American government during much of the twentieth century is a regular shift from civilian government to military regime. Sometimes little real difference has been apparent in policy, for the military has often been influential behind the scenes even during the periods of civilian rule. In addition, the political style of some military leaders has not been dissimilar to that of their civilian counterparts. Both kinds of leader have sometimes employed populist rhetoric to appeal to the masses. Both have relied on nationalist sentiment and have promised better living conditions for the poor. A few have managed to redistribute income during periods of economic prosperity only to fall from power when economic conditions deteriorated. Indeed, the economic situation seems to be the critical ingredient for success. Regimes of all kinds survive longer during economic booms than during recessions. During booms, any government can offer rewards in return for loyalty and support. During a recession, however, and especially when the economy is suffering from high rates of inflation, maintaining political stability is difficult. Frequent changes of government have usually been the consequence.

The military governments that came to power during the late 1960s and early 1970s, however, were rather different from earlier military regimes. These were not temporary interventions aiming to remove incompetent civilians and to replace them, after a short interlude, with another, more effective civilian administration. A new kind of military government seemed to take over, one that wanted power in order to change the nature of the economic and political system. The new regimes were highly committed to the capitalist path of development and were prepared to use harsh methods to counter opposition. Death squads, the abolition of human rights, and harsh economic policies were common, particularly in the so-called southern cone of Argentina, Brazil, Chile and Uruguay. Such governments also introduced supposedly more technically-based methods of administration, claiming to manage their countries more efficiently and effectively. Some of these governments survived for considerable periods; in Paraguay, Alfredo Stroessner's regime lasted for more than thirty years.

It is clear in practice, however, that few military regimes were able to solve the main economic problems of their countries. In some cases, they exacerbated the difficulties. Faced by mounting debt problems, rising rates of inflation, and widespread unpopularity they have gradually been in retreat. They have been replaced by civilian governments who are, in turn, faced by severe economic and social problems. No doubt the military will reappear in the not so distant future. If they do, however, it will not be a sign of the political immaturity of the Latin psyche, much more an indication of the appallingly difficult task of administering peripheral, capitalist economies. Rapid population growth, combined with a subordinant position in an unfairly structured world economy, does not make for easy government.

Key ideas

1 Social conditions improved in most Latin American countries up to 1980. Then the debt crisis led to falling living standards for poor and middle class alike.
2 Social discrimination is very marked in most societies. Race, language, income and gender are all important determinants of social position.
3 There is little sign of the distribution of income becoming more equal; in places, the economic model has made inequality worse.
4 Most countries have followed a capitalist development model, albeit with substantial amounts of state intervention. During the 1980s there has been a tendency to reduce state participation.
5 Poor peripheral societies are difficult to administer. Perhaps for that reason there have been many military coups in Latin America. Military regimes have alternated with civilian governments; in most cases neither has been able to resolve the basic economic and social contradictions inherent in Latin American development.

Further reading and review questions

Chapter 1

1 What were the principal legacies of colonial rule and to what extent do different countries still share those legacies today?
2 To what extent was growing 'dependence' linked to rising levels of economic development in Latin America?
3 How important is the physical background in explaining differences in the prosperity of the different regions of Latin America?

Further reading

Blakemore, H. and Smith, C. T. (1983) *Latin America: Geographical Perspectives*, London: Methuen, second edition.
Burbach, R. and Flynn, P. (1980) *Agribusiness in the Americas*, Monthly Review Press.
Gilbert, A. G. (1985) *An Unequal World*, Basingstoke: Macmillan Education.
Preston, D. (ed.) (1987) *Latin American Development: Geographical Perspectives*, Harlow: Longman, Chapter 2.
Stein, S. J. and Stein, B. H. (1970) *The Colonial Heritage of Latin America*, New York: Oxford University Press.

Chapter 2

1 Discuss the successes and failures of import substituting industrialization in any two countries.
2 In which Latin American countries is export-oriented industrialization most likely to be successful?

Further reading and review questions   93

3 In what ways did the policies of Latin American governments generate the debt crisis? To what extent was the debt crisis caused by outside events?
4 To what extent does the experience of different Latin American countries show that rapid population growth slows economic expansion?

Further reading

Gwynne, R. N. (1985) *Industrialisation and Urbanisation in Latin America*, London: Croom Helm.
Merrick, T. W., (1986) 'Population pressures in Latin America', *Population Bulletin* 41, Population Reference Bureau.
Roddick, J. (1988) *The Dance of the Millions: Latin America and the Debt Crisis*, London: Latin America Bureau.

Chapter 3

1 Under what circumstances is land reform both an equitable and an efficient strategy for rural development?
2 Who benefits and who loses from the commercialization of agriculture?
3 How might the strategy of colonization in Brazil have been improved?

Further reading

Lindqvist, S. (1979) *Land and Power in South America*, Harmondsworth: Penguin.
Preston, D. (ed.) (1980) *Environment, Society, and Rural Change in Latin America*, Chichester: John Wiley, Chapters 2–4.
Preston, D. (ed.) (1987) *Latin American Development*, Harlow: Longman, Chapter 8.

Chapter 4

1 What evidence is there to suggest that most cityward migrants make sensible decisions about where to live?
2 How might Latin American governments most effectively limit the growth of the largest cities?
3 Why has urban primacy developed in so many Latin American countries?

Further reading

Butterworth, D. and Chance, J. K. (1981) *Latin American Urbanization*, Cambridge: Cambridge University Press.
Gilbert, A. G. and Gugler, J. (1990) *Cities, Poverty and Development: Urbanization in the Third World*, Oxford: Oxford University Press.
Roberts, B. (1978) *Cities of Peasants*, London: Edward Arnold.
Stohr, W. and Taylor, D. (1981) *Development from Above or Below?*, Chichester: John Wiley.

Chapter 5

1 To what extent and in what ways are jobs in the informal sector similar?
2 How do large industrial plants benefit from the presence of the informal sector?
3 Under what circumstances is self-help construction an effective method of housing the poor?
4 Why are there relatively few urban protests in Latin America's cities?

Further reading

Bromley, R. and Gerry, C. (eds) (1979) *Casual work and poverty in Third World cities*, Chichester: John Wiley.
Gilbert, A. G. and Ward, P. M. (1985) *Housing, the State and the Poor: Policy and Practice in Latin American Cities*, Cambridge: Cambridge University Press.
Lloyd, P. (1981) *The Young Towns of Lima*, Cambridge: Cambridge University Press.
Perlmann, J. (1976) *Myth of Marginality*, Berkeley: University of California Press.
Ward, P. M. (ed.) (1982) *Self-help Housing: a Critique*, London: Mansell.

Chapter 6

1 Assuming that your parents were well educated, would you prefer to live in Brazil or in Cuba? Would you choose to live in Brazil or Cuba if your parents were poor farmers?
2 In what respects is life in Latin America harder for women than for men?
3 How feasible is the redistribution of income in Latin America in the absence of economic growth?
4 Why have there been so many military governments in the recent past in Latin America?

Further reading

Brundenius, C. (1981) 'Growth with equity: the Cuban experience (1959-1980)', *World Development* 9: 1,083-96.
Cordera, R. and Tello, C. (eds) (1984) *La Desigualdad en Mexico*, Siglo XXI.
Cubitt, T. (1988) *Latin American Society*, Harlow: Longman, Chapters 3, 4, 7 and 8.
Felix, D. (1983) 'Income distribution and quality of life in Latin America: patterns, trends and policy implications', *Latin American Research Review* 18: 3-34.
Open University (1983) *Industrialization and Energy in Brazil*, Third World Studies Case Study 6.
Portes, A., (1985) 'Latin American class structures: their composition and change during the last decades', *Latin American Research Review* 20: 7-39.
Turits, R. (1987) 'Trade, debt and the Cuban economy', *World Development* 15: 163-80.

Index

Agriculture:
  colonial 30–1
  commercialization 38–43
  encomienda 2–3
  export 8–9, 19–20
  labour 16
  production 40–1
Allende, Salvador 34, 86
Arbenz, Jacobo 32
Argentina:
  agriculture 8–9, 15
  debt 26–8
  education 15, 76, 80
  exports 8–9, 19
  income 15, 29, 77
  income, distribution of 82
  independence 6–8
  industry 21, 22, 23, 64
  military rule 90
  population 16–17
  public sector 90
  urban growth 45
  urban primacy 51
Asunción 54
Aztecs 2

Barranquilla 52, 61, 66
Belo Horizonte 52, 53
Bogotá 45, 52, 53, 56, 61, 66, 67, 78
Bolívar, Simón 6
Bolivia:
  debt 26–8
  education 15, 76, 80
  exports 9, 19–20
  income 15, 29, 77
  independence 6–8
  industry 22
  land reform 32, 33
  population 16–17
  urban growth 45
  urban primacy 51
Brasília 36–7, 54–5, 59, 66, 71
Brazil:
  agriculture 15, 36–8, 40, 42
  Amazonia 35, 36–8, 42
  colony 3, 30, 31
  debt 26–8, 88–9
  economic 'miracle' 88–9
  education 15, 76, 80
  exports 8, 18–19, 24, 88–9

income 15, 88
income, distribution of 82–3
independence 6–8
industry 20, 21, 22, 23, 62, 64, 87, 88–9
land colonization 35, 36–8, 42
military rule 88–9, 90
population 16–17
public sector 89
race and ethnicity 78–9
slavery 31
urban growth 45
urban primacy 51
Buenos Aires 44, 50–1, 52, 53, 56, 59, 61, 66, 67, 70–1, 78

Caracas 45, 49–50, 52, 53, 54–5, 59, 60–1, 66, 71
Cárdenas, Lázaro 32
Castro, Fidel 84
Catholic Church 3, 4, 17, 31
Central America:
  Common Market 24
  debt 26–8
  education 76, 80
  exports 19, 38
  income, distribution of 82
  independence 7–8
  industry 22
  land reform 32, 33
  urban growth 45
Chile:
  agriculture 15
  debt 26–8
  education 15, 76, 80
  exports 19
  income 15, 77
  independence 6–8
  industry 21, 22
  land reform 32, 33, 34–5
  military rule 90
  population 16–17
  public sector 89
  urban growth 45
  urban primacy 51
Ciudad Guayana 54–5, 60–1
Ciudad Lázaro Cárdenas 54–5
climate 10–12

coffee 12, 18–20, 31
Colombia:
  debt 26–8
  education 15, 76, 80
  exports 8, 18–19, 24
  independence 6–8
  income 15, 77
  migration 47–56
  population 16–17
  urban growth 45, 53–4
  urban primacy 51
Colonial rule:
  agricultural system 30–1
  independence 6–8
  Portuguese conquest 3
  Spanish conquest 1
Córdoba 52
Cortés, Hernán 3
Columbus, Christopher 1
Costa Rica 7, 8, 19, 24, 82
Cuba:
  agrarian reform 39, 84
  debt 26–8
  development model 84–6
  education 15, 76, 80, 85
  exports 19, 84–5
  gender 81
  income 77, 84
  income, distribution of 83
  independence 6–8, 84
  land holding 39
  population 16–17, 85
  slavery 31
  Soviet Union 84–6
  urban growth 45
  urban primacy 51, 55, 58
Curitiba 52
Cuzco 2

Debt crisis 25–9
'dependence' 9–10, 20–3, 40–2, 84, 86
Dominican Republic:
  debt 26–8
  education 15, 76, 80
  exports 19, 24
  income 15, 77
  independence 6–8

industry 22
population 16–17
urban growth 45

earthquakes 11–12
Ecuador:
  agriculture 15
  debt 26–8
  education 15, 76, 80
  exports 19, 20
  income 15
  independence 6, 8
  industry 22
  population 16–17
  urban growth 45
  urban primacy 51
Education 76, 80, 85
El Salvador 7, 8, 19, 24, 32, 82
employment:
  female participation 81–2
  formal sector 61–4
  informal sector 61, 64–5, 72–3
  structure of 15
  unemployment 77–8, 85
  urban 60–5

foreign trade:
  exports 5–6, 8–9, 18–20, 24–5, 27
  free trade areas 24
  imports 5–6, 8–9, 21
Fortaleza 52

gender 81–2
Guadalajara 52, 53, 54–5, 59, 60–1, 66, 71
Guatemala 2, 4, 7, 8, 15, 16, 17, 19, 24, 25, 27, 32, 33, 45, 51, 76, 77
Guayaquil 52

Havana 52, 53, 55–8
Honduras 7, 8, 19, 24
housing:
  rental 66, 68–70, 74
  self-help 65–70, 74
hurricanes 12

Incas 2
income, distribution of 82–6
industrialization:
  export orientation 23–5
  import substitution 21
  steel 21
infrastructure and services 66, 69–70, 71, 73, 74, 76

Juárez, Benito 78

Kubitschek, Juscelino 57

land:
  colonization 31, 35–8
  reform 32–5
  tenure 2–3, 30–2, 42
  urban 68–9, 74
language 4, 79
Lima 52, 53, 66, 77, 78
living standards 14–15, 29, 76–8

Maracaibo 66
Mayas 2, 4
Medellín 52, 53, 61, 66, 78
Mexicali 49
Mexico:
  agriculture 15, 40, 42–3
  debt 26–8
  education 15, 76, 80
  exports 19–20, 24
  income 15, 77
  income, distribution of 82–3
  independence 6–8
  industry 21, 22, 23, 62, 64
  land colonization 35
  land reform 32, 33
  maquiladoras 24
  migration 47–9
  oil 20, 26
  population 16–17
  public sector 89
  race 78
  revolution 33
  urban growth 45
  urban primacy 51
Mexico City 2, 4, 11, 44, 52, 53, 59, 63, 69, 70, 78

migration 45–50
military rule 88–9, 89–91
Monterrey 52, 53
Montevideo 52, 78

Nicaragua 7, 8, 11, 19, 24, 32, 33, 86

oil 19–20, 26, 51, 88

Panama 7, 8, 10, 19
Paraguay:
  exports 19
  independence 6–8
  military rule 91
  urban growth 44
Peru:
  agriculture 15
  debt 26–8
  education 15, 76, 80
  exports 19–20
  income 15, 29, 77
  income, distribution of 82
  independence 6–8
  industry 21, 22
  land reform 32, 35
  population 16–17
  urban growth 45
  urban primacy 51
Pizarro, Francisco 3
Pinochet, Augusto 34
politics 70–4
population:
  family planning 17–18, 85
  growth 13–18
  pre-Columbian 1–2
  racial mixing 4–5
Pôrto Alegre 52
primacy, urban 50–4
privatization 89

race and ethnicity:
  discrimination 78–9
  gender 81
  miscegenation 4–5
Recife 52
relief 10–12
religion 3, 4, 17, 30

Rio de Janeiro 8, 19, 44, 52, 53, 54, 62, 66, 78
Rosario 52
rural development 30–43

Salvador 3, 52
Santiago (Chile) 52, 53, 78
San Martín, José de 6
Santo Domingo 52
São Paulo 8, 42, 52, 53, 59, 69
social class 6, 79–80
social indicators 14–15, 76–8
slavery 3, 31
socialism 85–6, 86–7
Stroessner, Alfredo 91

Tijuana 49, 66, 69
transnational corporations 21–2, 24, 38–40

unemployment 77–8, 85
urban deconcentration 54–9
urban development 44–5
urban primacy 50–4
Uruguay:
  exports 18–19
  independence 6–8, 10
  military rule 90

Valencia 74
Venezuela:
  agriculture 15
  debt 26–8
  education 15, 76, 80
  exports 8–9, 19–20
  income 15, 77
  income, distribution of 82
  independence 6–8
  industry 21, 22, 62, 64
  land reform 35
  migration 49–50
  oil 20
  population 16–17
  urban growth 45, 54–55, 60–1
  urban primacy 52
Viedma 54–5

wages 77